BEAUTY &
REVOLUTION
IN SCIENCE

REVOLUTION
IN SCIENCE

Cornell University Press   ITHACA AND LONDON

First published 1996 by Cornell University Press.

Printed in the United States of America

⊗ The paper in this book meets the minimum requirements
of the American National Standard for Information Sciences—
Permanence of Paper for Printed Library Materials, ANSI Z39.48—1984.

Library of Congress Cataloging-in-Publication Data

McAllister, James W. (James William), 1962–
Beauty and revolution in science / James W. McAllister.
p.   cm.
Includes bibliographical references and index.
ISBN 0-8014-3240-5 (alk. paper)
1. Science—Philosophy.   2. Science—Mathematical models.
3. Aesthetics.   4. Rationalism.   I. Title.
Q175.M415   1996
501—dc20   96-3910

Contents

Contents

Contents

Illustrations

Acknowledgments

For their comments on early versions of the arguments presented here, I thank Professor Mary B. Hesse and Professor Nicholas Jardine, University of Cambridge, and Professor James R. Brown, University of Toronto. They do not entirely agree with my views, as will be obvious from their own writings. I am grateful to two anonymous referees of Cornell University Press for their perceptive comments on the penultimate draft. Lastly, I thank my colleagues and friends at the University of Leiden for providing the pleasant environment in which this work was completed.

Some material in this book is developed from the following publications: "Truth and Beauty in Scientific Reason," in *Synthese* 78 (1989), 25–51 (© 1989 by Kluwer Academic Publishers); "The Simplicity of Theories: Its Degree and Form," in *Journal for General Philosophy of Science* 22 (1991), 1–14 (© 1991 Kluwer Academic Publishers); "Scientific Realism and the Criteria for Theory-Choice," in *Erkenntnis* 38 (1993), 203–222 (© 1993 Kluwer Academic Publishers); and "Scientists' Aesthetic Preferences among Theories: Conservative Factors in Revolutionary Crises," in Alfred I. Tauber, ed., *The Elusive Synthesis: Aesthetics and Science* (Dordrecht: Kluwer, 1996), 169–187 (© 1996 Kluwer Academic Publishers); all used by permission of Kluwer Academic Publishers; "Scientists' Aesthetic Judgements," in *British Journal of Aesthetics* 31 (1991), 332–341, by permission of Oxford University Press; "The Formation of Styles: Science and the Applied Arts", in Caroline A. van Eck, James W. McAllister, and Renée van de Vall, eds., *The Question of Style in Philosophy and the Arts* (Cambridge: Cambridge University Press, 1995), 157–176, by permission of Cambridge University Press.

J. W. M.

BEAUTY &
REVOLUTION
IN SCIENCE

Introduction

Ever since what we know as science first arose, philosophers have striven to describe and understand scientific practice by constructing models of it. Scientific practice shows great variety, however: it takes different forms in different branches of science, historical periods, research schools, and individual scientists. No unified model is yet available that accounts for scientific practice in all this variety. As long as such a model eludes us, the best way to describe and understand scientific practice is to construct various partial models, each of which accounts reasonably well for one or another facet of the subject matter. For this reason, philosophy of science abounds with partial models of scientific practice.

We can conceive of these partial models as arranged on levels corresponding to their breadth of scope. Models on the highest level, such as falsificationism and inductivism, aim to account for the broadest features of scientists' work or the largest-scale developments in the history of science, but are insufficiently articulated to explain more detailed features of scientific practice, such as scientists' resistance to new theories or their recourse to thought experiments. Models on intermediate levels, such as accounts of analogical reasoning, shed light on individual methodological devices but do not presume to describe every instance of theory succession. Models on yet lower levels, which chronicle particular periods in the development of a science, may show excellent accord with historical evidence but do not lend themselves to generalization. At the lowest levels are found items of scientists' autobiography: their occasional reflections about the problems on which they have worked and the approaches that they have used.

The logical relations among these models are intricate. Many of the models occupying the highest level, like falsificationism and inductivism, contradict one another and must therefore be regarded as rivals. Models at the lowest levels may conflict with each other too, but more typically they treat distinct historical episodes and are therefore logically independent of one another. Each of the highest-level models is consistent with some lower-level models and typically with more and more numerous models at lower and lower levels: very many items of scientists' testimony are consistent with falsificationism, for instance. It is therefore possible to arrange a selection of partial models of science into a pyramidal structure containing one top-level model, several medium-level accounts, and many low-level models. A well-formed pyramid of models will offer its user an understanding of features of science on all scales, from the broadest sweep to the smallest detail. Each philosopher of science holds explicitly or implicitly to such a pyramid of models, which provides his or her view of scientific practice. Much of the debate in philosophy of science is occupied with comparing the merits of alternative pyramids of models.

This book is a contribution to what I regard as the most convincing of the pyramids of models about science presently available. The top level of this pyramid is occupied by the model that I shall call "the rationalist image" of science. The rationalist image holds that there exists a set of precepts for investigating and reasoning about the world that have a privileged relationship with reality: the precepts of rationality. The rationalist image commits its adherents to providing rationalist accounts of all features of scientific practice, though of course not to describing all scientists' acts as rational. This book contributes to the pyramid of models headed by the rationalist image by constructing a rationalist model of two features of scientific practice that have so far evaded explanation on rationalist principles: the appeal that scientists make to aesthetic criteria in evaluating their theories, and scientific revolutions.

The model that I present in this book is a medium-level model of scientific practice, of a scope intermediate between the loftiest generalization and the historical case study. Models on this level can match neither the peremptory simplicity of top-level models nor the detail and sensitivity of historical studies of individual episodes. The latter is certainly the more serious limitation, and one that I make no attempt to conceal in the model that I offer. I invite anyone who wishes to obtain the finest-grained picture of particular episodes in the history of science, such as the rise of heliocentrism in astronomy or of quantum theory in physics, to look elsewhere. Here we deal at a somewhat higher level of generality, searching for the elements that are common to classes of historical episodes and accepting the loss of detail that this entails.

Introduction

The plan is as follows. Chapter 1, "Two Challenges to Rationalism," points out that rationalists have for some decades met difficulty both in explaining why scientists should make such extensive appeal to aesthetic considerations in theory evaluation as they do, and in giving a convincing account of scientific revolutions. The aim of this book is to remedy this deficiency of rationalist accounts of science. Chapter 2, "Abstract Entities and Aesthetic Evaluations," presents the conceptual apparatus that we will employ in this task. Throughout this book, our attention will be directed at the aesthetic properties of scientific theories themselves, which are abstract entities, and not at the properties of representations of theories in concrete form, such as in texts and diagrams. Chapter 2 draws this distinction and conducts a brief polemic against one nonrationalist view of science, the actor-network theory, that neglects the concept of scientific theory in favor of that of inscription. Further, this chapter portrays scientists as holding to aesthetic criteria, each of which attributes aesthetic value to a particular property of theories.

Chapter 3, "The Aesthetic Properties of Scientific Theories," surveys some of the properties of theories to which scientists have attached aesthetic value. I group the aesthetic properties that theories may show into classes: for example, one such class comprises the various symmetry properties that theories can show. This survey provides evidence that scientific communities perform two sorts of evaluations of theories: one is directed at ascertaining the theories' likely empirical performance, whereas the other employs terms of aesthetic appreciation.

What is the relation between evaluations of these two sorts? A spectrum of possible answers can be envisaged, each claiming that aesthetic judgments are reducible to a particular extent to empirical judgments. At one extreme of this spectrum lies the claim that scientists' aesthetic evaluations are disinterested about the empirical virtues of theories, so that scientists' aesthetic and empirical evaluations of theories are independent of one another. If this claim were correct, one would expect to find in the historical record no systematic correlation between the aesthetic and empirical verdicts that scientists have actually passed on theories. At the other extreme of the spectrum is the view that scientists' aesthetic judgments and their empirical judgments are nothing but manifestations or aspects of one another. Two forms of this view may be envisaged: the first portrays aesthetic judgment as an aspect of empirical judgment, while the second reduces empirical judgment to aesthetic judgment. In either case, the aesthetic and empirical verdicts that scientists pass on theories would always necessarily agree.

These extreme views are discussed in Chapter 4, "Two Erroneous Views of Scientists' Aesthetic Judgments." I give reasons, in the form mainly of evidence from scientists' practice, for rejecting each of them as a model of how scientists reach their aesthetic evaluations of theories. In Chapter 5, "The Inductive Construction of Aesthetic Preference," I present a third model as superior. According to this new model, a scientific community's aesthetic preferences are reached by an induction over the empirical track record of theories: a community attaches to each property of theories a degree of aesthetic value proportional to the degree of empirical success of the theories that have exhibited that property. I call this procedure the "aesthetic induction."

On my view, we have no guarantee that there is a correlation between particular aesthetic properties and high degrees of empirical adequacy in theories. Like all policies of inductive projection, however, the aesthetic induction can be expected—provided that it is pursued for long enough—to discern any such correlation that may exist. We examine in Chapter 6, "The Relation of Beauty to Truth," the possibility that the aesthetic induction may discern such a correlation in the course of the history of science. Many twentieth-century scientists, including Albert Einstein, seem to have concluded that such a correlation has already been found, but we shall see that the evidence does not support this conclusion.

Scientists frequently judge theories for the simplicity properties that they exhibit, and philosophers of science have devoted much discussion to this practice. No agreement has so far been reached about the extent to which scientists' simplicity considerations are empirical or aesthetic. In a reexamination of this issue in Chapter 7, "A Study of Simplicity," I suggest that scientists in fact appeal to two separate simplicity criteria: one that attaches value to a particular form of simplicity, and one that (usually) favors theories in which this form is shown to a higher degree. Whereas there may be some justification for the latter criterion on empirical grounds, I suggest that the former is an aesthetic criterion, which is periodically updated by inductive projection. This means that if there is some form of simplicity that is strongly correlated with empirical adequacy, the scientific community is capable of identifying it, provided that science is pursued for long enough.

The aesthetic induction explains how scientists' criteria for theory choice evolve gradually, but it is unable on its own to account for scientific revolutions, in which standards for theory choice change suddenly. Chapter 8, "Revolution as Aesthetic Rupture," shows how the model of scientific practice so far constructed can be extended to describe revolutions. Consider the sequence of empirically successful theories adopted by a scientific community. If each of these theories shows aesthetic properties similar to those of its predecessors, the aesthetic induction will be able to revise the community's aesthetic canon sufficiently promptly that the community's empirical and aesthetic criteria in cases of theory choice will tend to agree. But if theories in the sequence suddenly come to show aesthetic properties that are unprecedented, the aesthetic induction may be unable to revise the aesthetic canon sufficiently quickly to reflect this development. The community's aesthetic criteria will therefore enter into conflict with its empirical criteria. I interpret a scientific revolution as the rupture with an aesthetically defined tradition that empirically minded scientists in such circumstances perform.

Lest my claim that scientists' aesthetic preferences are shaped by utilitarian concerns and through inductive projection should seem implausible, I present in Chapter 9, "Induction and Revolution in the Applied Arts," a view of the formation of styles in the applied arts. Design in the applied arts is constrained both by the technical means available and by the aesthetic canons that are in force. A new material cannot be exploited to the full in structures unless their design responds to its characteristics. But the first designs that exploit a new material in a substantial way frequently strike onlookers as aesthetically unappealing, since the aesthetic canons that predominate at any time are generally tailored to the peculiarities of longer-established technical means. I show that the aesthetic canons by which designs are appraised in the applied arts evolve in response to utilitarian concerns: the community comes to value designs in which technical innovations can be exploited most thoroughly. There are close parallels between this process in the applied arts and the phenomena that we have been discussing in scientific practice. From these parallels I draw two conclusions: first, it does not conflict with our understanding of aesthetic canons in the arts to suggest that scientists' aesthetic preferences are shaped by inductive projection over the perceived empirical performance of their theories; second, aesthetic preferences in practices as different as the sciences and the applied arts are shaped partly by habituation to the forms associated with success.

Chapters 10, "Circles and Ellipses in Astronomy," and 11, "Continuity and Revolution in Twentieth-Century Physics," contain case studies that display the power of this model of scientific practice to account for historical episodes. Two pairs of episodes are discussed: the rise of Copernicus's theory and Kepler's theory in mathematical astronomy, and the rise of relativity theory and quantum theory in physics. Each of these four episodes is frequently portrayed as a revolution, but I shall argue that only the second of each pair should be considered revolutionary.

The final chapter, "Rational Reasons for Aesthetic Choices," returns to the two challenges to the rationalist image of science. We examine anew to what extent scientists' practice of evaluating theories for their aesthetic properties has a rational justification, and to what extent the occurrence of revolutions shows that there is no such thing as scientific rationality. I aim to show that, contrary to the fears of some philosophers and the hopes of others, the rationalist image of science is not undermined by either scientists' appeals to aesthetic considerations or their participation in revolutions.

Two Challenges to Rationalism

One can always make a theory, many theories, to account for known facts, occasionally even to predict new ones. The test is aesthetic.
—George Thomson, *The Inspiration of Science*

1. The Rationalist Image of Science

According to the rationalist image of science, there exists a set of precepts for conducting science—the norms of rationality—which admits of some principled and extrahistorical justification. There is, in other words, a basis for forming and judging decisions and policies in scientific practice that does not depend on convention, fashion, or other local or historical phenomena. A subsidiary claim made by the rationalist image of science is that, while individual decisions and policies of scientists in history may deviate from those that would have been advised on rational grounds, such deviations have not been excessively wide or persistent: actual science is predominantly rational. As many philosophers of science have noted, the rationalist image is a convincing high-level model of scientific practice: it accounts persuasively for much behavior of scientists and for many episodes of the history of science.[1]

In recent decades, however, two bodies of historical evidence have emerged that have led some philosophers and historians of science to question the adequacy of the rationalist image of science. The first of these establishes that the development of science is punctuated by revolutions, events in which a community's norms for the formulation and assessment of theories change radically. The second body of evidence establishes that scientists make substantial and systematic appeal to aes-

---

1. I regard what I call the rationalist image as the model of science underlying such works as Popper (1959), Lakatos (1970), Laudan (1977), and Newton-Smith (1981).

thetic preferences in judging available theories and in choosing between them. We shall be reviewing this evidence in detail at the appropriate junctures.[2]

These bodies of evidence weigh against the rationalist image in the following manner. Take first the occurrence of revolutions. The model of such events that has so far had the most influence among philosophers and historians, that of Thomas S. Kuhn, claims on some interpretations that revolutions consist of a change of the community's criteria for theory assessment in their entirety: no methodological precept survives a revolution unaltered. This means that there is no set of methodological precepts which retains validity throughout the history of science and therefore that there can exist no canon of rationality. Supporters of this view would regard the phrase "canon of rationality" as, at most, a synonym for "style of reasoning"—a label to be applied to whichever set of basic methodological precepts is obeyed at a particular time in a community.[3] A similar conclusion is reached by many of those who remark on the incidence of aesthetic judgments in science. Most people regard aesthetic preferences as irremediably emotive and idiosyncratic, and consequently presume scientists' aesthetic preferences to be unrelated to empirical adequacy or to any other rationally desirable property of theories. On this view, for scientists to rely on aesthetic criteria in judging theories is irrational. This view of scientists' aesthetic preferences is put forward for instance by Helge Kragh:

> The principle of mathematical beauty, like related aesthetic principles, is problematical. The main problem is that beauty is essentially subjective and hence cannot serve as a commonly defined tool for guiding or evaluating science. It is, to say the least, difficult to justify aesthetic judgment by rational arguments. [. . .] I, at any rate, can see no escape from the conclusion that aesthetic judgment in science is rooted in subjective and social factors. The sense of aesthetic standards is part of the socialization that scientists acquire; but scientists, as well as scientific communities, may have widely different ideas of how to judge the aesthetic merit of a particular theory. No wonder that eminent physicists do not agree on which theories are beautiful and which are ugly.[4]

2. Evidence for believing that the development of science is punctuated by revolutions is surveyed in I. B. Cohen (1985), pp. 40–47. Previous books on aesthetic factors in science are Wechsler (1978), Curtin (1982), Chandrasekhar (1987), Rescher (1990), and Tauber (1996), though only parts of these discuss the role of aesthetic considerations in the evaluation of theories. Alexenberg (1981), pp. 146–202, interviews scientists on aesthetic experiences that they have undergone in their work.

3. Kuhn (1962).

4. Kragh (1990), pp. 287–288.

If this view of scientists' aesthetic preferences is correct, the progression of science, which is no more than the outcome of a sequence of acts of theory choice, is influenced systematically and substantially by irrational factors.

This book aims to defuse the threat posed to the rationalist image of science by these two bodies of historical evidence. I shall try to show that both the evidence of the occurrence of revolutions and the evidence of scientists' appeal to aesthetic considerations are consistent with the rationalist image. The intended outcome of my treatment is a rationalist view of science that allows us to accept both that scientific method has undergone radical and sudden transformations, and that aesthetic considerations are among the grounds on which scientific communities choose between competing theories.

On the account that I shall offer, these two phenomena of science are closely interrelated. Indeed, the occurrence of scientific revolutions is a consequence of scientists' use of aesthetic criteria for theory evaluation. If this is so, the key to a rationalist understanding of scientific revolutions lies in scientists' aesthetic preferences. The bulk of the book will thus be devoted to this second topic: we shall return to revolutions in Chapter 8.

2. A Rationalist Model of Theory Evaluation

The component of the rationalist image brought most directly into question by the historical evidence about the occurrence of revolutions and the incidence of aesthetic judgments is its account of scientists' evaluations of their theories. We should therefore begin our defense of the rationalist image by recalling how rationalists view the practice of theory assessment in science. There are of course several alternative models of this practice that a rationalist may advance: here I pick one that has been set out by W. H. Newton-Smith, and which I shall call the "logico-empirical model" of theory assessment.[5]

This model is based on the following premises. Science's ultimate goal is the production of the most complete and accurate account possible of the universe. Theories approximate to this ideal to the degree to which they possess the property "empirical adequacy." The statement that a theory has empirical adequacy to the highest degree possible means that its claims are true of all observable phenomena, including phenomena in the past and phenomena in other ways inaccessible to us; the statement that a theory has empirical adequacy to a somewhat lesser degree means

5. Newton-Smith (1981), pp. 208–236.

that its claims are true of a similar proportion of observable phenomena. Scientific realists, who would say that the ultimate goal of science is the production of an account of the universe that is true, can nonetheless concur with this analysis, since they see the degree of empirical adequacy of a theory as a consequence of its being to a corresponding degree close to the truth.[6]

It might initially seem that the only criterion for theory assessment that the logico-empirical model need recommend is a criterion of empirical adequacy itself: "Prefer a theory that has a higher degree of empirical adequacy to one that has a lower degree of it." However, the meaning of "empirical adequacy" makes it impossible to use this criterion in practical choices among theories. The only way in which we could establish that a theory possesses empirical adequacy to the highest degree possible would be to demonstrate that it accords with all empirical data that could be gathered from all sources over unlimited time spans; similarly, we could establish that a theory has a particular lesser degree of empirical adequacy only by showing that it accords with a corresponding proportion of those data. Obtaining a direct reading of the degree of a theory's empirical adequacy would therefore involve ascertaining the proportion of all empirical data with which the theory accords. But, even if the notion of counting and comparing the number of the confirmed and disconfirmed predictions of a theory could be made precise, such a task cannot be completed in a finite time for generalizations of wide scope other than tautologies or contradictions. Thus, the criterion of empirical adequacy itself does not provide a practical basis for choosing among competing theories.[7]

We can, however, identify other criteria that are diagnostic of high degrees of empirical adequacy in theories and that yield their verdicts quickly enough to be useful in theory evaluation. We may construct a set of such criteria by considering what properties a theory must possess if it is to have a high degree of empirical adequacy: it should exhibit accord with a high proportion of the phenomena investigated hitherto and show some promise of according with a high proportion of phenomena not yet studied. On this basis, the logico-empirical model prescribes criteria such as the following:

1. The criterion of consistency with extant empirical data: other cir-

6. For the agreement of both instrumentalists and scientific realists with the claim that science aims at theories that have empirical adequacy to high degree, see van Fraassen (1980), p. 12, and Churchland (1985), pp. 38–39.
7. I have investigated these consequences of the meaning of "empirical adequacy" for theory assessment in McAllister (1993).

cumstances being equal, a theory should be more highly valued if its implications agree with what is now known of phenomena.

2. The criterion of novel prediction: a theory should further be valued if it offers predictions of, and subsequently accords with, data that were not available when the theory was formulated, or at least that were not taken into account in its formulation. After all, if the sole empirical requirement of theories were that they should accord with data gathered previously, a theory constructed deliberately to account for available data would have to be given a high score; and it is possible in any circumstance to construct infinitely many such theories.

3. The criterion of consistency with current well-corroborated theories: a new theory should be more highly valued if, other circumstances being equal, it coheres with other theories that received high scores on the previous criteria. As a supporter of scientific realism would argue, a set of true theories about the world would all be consistent with one another; so, if we now have any theories that we think are close to the truth, we should wish any new theory that we adopt to be consistent with them.

4. The criterion of explanatory power: while a new theory is minimally required not to contradict well-established theories, it should be more strongly valued if it can provide an explanation of the generalizations that they contain. Such an attainment suggests that the theory has identified a pattern or mechanism underlying the data, and it offers a prospect that the theory will accord with sets of data yet to be gathered.

An addition to this list of criteria is made necessary by the following consideration. If all that we wanted from science were theories that are logically compatible with data, we would be satisfied with theories that are tautologies and logical contradictions. After all, there is no logically possible state of affairs that a tautology rules out, and any prediction whatever can be derived from a contradiction. But such statements cannot be regarded as having high degrees of empirical adequacy, as their predictions are not determinate: they do not allow us to distinguish the universe that we inhabit from all other logically possible universes. In order to prevent our empirical criteria for theory choice from leading us to embrace tautologies and contradictions, the logico-empirical model must add to the above list two further criteria:

5. The criterion of empirical content: theories must not be tautologies.

6. The criterion of internal consistency: theories must not contain internal contradictions.

The logico-empirical model of theory assessment has the task of accounting on rationalist principles for scientists' preferences among theories. It discharges this task well: very many choices among theories that scientists perform can be explained by supposing that they are decided on criteria such as the six listed above. The logico-empirical model of theory assessment is thus a valuable extension of the rationalist image of science.

This model fails, however, to provide the rationalist image with a satisfactory response to the two bodies of historical evidence whose challenge we are examining. Consider first the evidence that science undergoes revolutions. How can the logico-empirical model account for this fact? The logico-empirical model's six criteria for theory assessment listed above are derived exclusively from an analysis of "empirical adequacy." Therefore, if these criteria are valid, they must be valid at all times, unless the goal of science changes. This ensures that the logico-empirical model has no resources to explain how a scientist at one time can hold to criteria for theory assessment different from those of scientists of any other time. But scientific revolutions are episodes in which scientists' criteria for theory assessment change: so the logico-empirical model is incapable of explaining revolutions.

Similarly, the logico-empirical model is unable to make sense of the evidence that scientists appeal to aesthetic criteria in deciding among theories. Being couched entirely in terms of logical and empirical concerns, it lacks the apparatus to analyze aesthetic preferences. If the aesthetic predispositions of scientists are as idiosyncratic and irreducible to rational deliberation as many suppose them to be, then theory succession could hardly follow the path that the logico-empirical model prescribes.

I conclude that the logico-empirical model of theory assessment is not sufficiently sophisticated to account for the evidence of either the occurrence of revolutions or the incidence of aesthetic considerations in theory choice. Of course, it is open to the logico-empirical model to dismiss aspects of scientific practice by calling them irrational; but this option amounts to declaring parts of scientific practice inexplicable, to which rationalists should resort only in localized and exceptional cases. The rationalist image can meet the challenge posed by the evidence of the occurrence of revolutions and the incidence of aesthetic considerations in theory choice, but only if provided with a richer model of scientists' preferences among theories.

3. AESTHETIC FACTORS IN DISCOVERY AND JUSTIFICATION

In this section and the next, we examine two past attempts made by rationalists to dismiss the evidence of scientists' use of aesthetic criteria in theory choice. If either of these attempts had succeeded, rationalism would not be troubled by the need to account for scientists' aesthetic preferences among theories. Unfortunately, for reasons that will become clear, both fail.

The first attempt was made by logical positivism, a brand of rationalism that rose to prominence in the 1920s and long remained influential. Logical positivists advanced the thesis that a scientist working on a theory successively enters two "contexts." First is the "context of discovery," in which the scientist originates the theory by means of intuitions or conjectures. These acts are not guided by precepts of logic or rationality and therefore cannot be analyzed within a rationalist framework: there can be no logic of scientific discovery but only a psychology of it. Thereafter the scientist enters the "context of justification," in which he or she tests the theories that have been originated in the context of discovery. This testing occurs on logical and empirical criteria, and assures the rationality of theory succession.[8]

Logical positivists conceded that aesthetic factors could affect a scientist's behavior in the context of discovery, since they thought that a scientist could be inspired to formulate a hypothesis by a stimulus of any sort. But they rejected the suggestion that aesthetic factors played any part in the context of justification, presumably because they could conceive of no way in which aesthetic criteria could be assimilated to logical or empirical criteria. This attitude toward aesthetic factors in science is expressed by Herbert Feigl:

> A few words on some misinterpretations stemming from predominant concern with the history and especially the *psychology* of scientific knowledge. In the commendable (but possibly utopian) endeavor to bring the "two cultures" closer together (or to bridge the "cleavage in our culture") the more tender-minded thinkers have stressed how much the sciences and the arts have in common. The "bridges" [. . .] are passable only in regard to the *psychological* aspects of scientific [. . .] creation [. . .]. Certainly, there are esthetic aspects of science [. . .]. But [. . .] what is primary in the appraisal of scientific knowledge claims is (at best) secondary in the evaluation of works of art—and vice versa.[9]

According to logical positivism, therefore, there exists no such phenomenon as scientists' aesthetic evaluation of their theories and therefore no such phenomenon that need trouble philosophers of science. It is possi-

---

8. The origin and reception of the distinction between contexts of discovery and of justification are studied by Hoyningen-Huene (1987).

9. Feigl (1970), pp. 9–10.

Two Challenges to Rationalism

ble that scientists are affected by aesthetic factors in discovery, but drawing up an account of that phenomenon would be a task for biographers and psychologists of scientists rather than for philosophers of science.

Logical positivism has generally been superseded within philosophy of science, but it still overshadows the discussion of the role of aesthetic factors in science. The view persists that whereas aesthetic factors may be important in the creation of a theory, only empirical criteria can play a role in its acceptance. For example, Dean K. Simonton writes: "No scientist, including Dirac, would ever be so bold as to justify a theory on so irrational a basis as 'beauty.' "[10]

Logical positivists were undoubtedly correct in reporting an incidence of aesthetic considerations in the context of discovery: it frequently happens that a scientist picks the theories on which he or she will work in part on the strength of their aesthetic properties.[11] But in denying that aesthetic considerations play a part in scientists' assessments of theories, logical positivists neglected two facts. First, it is possible to regard intellectual creations of many kinds, ranging from mathematical proofs to chess games, as works of art. When we consider intellectual creations in this manner, we are led to appraise them for their aesthetic properties, and this aesthetic appraisal affects our overall view of and regard for them. It would be unusual if scientists were not tempted sometimes to regard scientific theories as works of art and to allow their overall view of them to be affected by aesthetic judgment. Of course, scientists frequently do surrender to both temptations. Ernest Rutherford, speaking in 1932, offers an example of this tendency:

> I think that a strong claim can be made that the process of scientific discovery may be regarded as a form of art. This is best seen in the theoretical aspects of Physical Science. The mathematical theorist builds up on certain assumptions and according to well understood logical rules, step by step, a stately edifice, while his imaginative power brings out clearly the hidden relations between its parts. A well constructed theory is in some respects undoubtedly an artistic production. A fine example is the famous Kinetic Theory of Maxwell. [. . .] The theory of relativity by Einstein, quite apart from any question of its validity, cannot but be regarded as a magnificent work of art.[12]

10. Simonton (1988), p. 193. For another recent denial that aesthetic factors play an important role in theory justification, see Engler (1990), p. 31.

11. Some comments on the heuristic role of aesthetic factors in science are to be found in Mamchur (1987).

12. Quoted from Badash (1987), p. 352. A discussion of the incidence of aesthetic factors in both the pursuit and the justification of theories is given by Chandrasekhar (1989).

Second, logical positivists omitted to recognize that scientists in their own work do not distinguish sharply between a context of discovery and one of justification. In most cases, the factors that lead a scientist to formulate a theory having certain properties also play a role in shaping the community's opinion about that theory's worth. In particular, it appears that scientists appeal to aesthetic factors both in their efforts to originate hypotheses and in their evaluations of theories that have been proposed in their community. By dismissing scientists' aesthetic evaluations of their theories as unimportant, logical positivists fail to render justice to this aspect of scientific practice.

The discrepancy between scientists' actual uses of aesthetic considerations and the logical positivist account of them is revealed by the writings of P. A. M. Dirac. In his many reflections on the role of aesthetic factors in his own work and in scientific practice generally, Dirac stressed their influence both as heuristic guides and as grounds for theory evaluation. First, as he admitted, Dirac used aesthetic criteria to decide priorities in his own research.[13] He thought that many of his colleagues worked in the same way. For instance:

> When Einstein was working on building up his theory of gravitation he was not trying to account for some results of observations. Far from it. His entire procedure was to search for a beautiful theory [. . .]. Somehow he got the idea of connecting gravitation with the curvature of space. He was able to develop a mathematical scheme incorporating this idea. He was guided only by consideration of the beauty of these equations. [. . .] The result of such a procedure is a theory of great simplicity and elegance in its basic ideas.[14]

Second, Dirac relied on aesthetic criteria also in assessing theories. "Context of discovery" and "context of justification" merge indissolubly in such statements as the following: "It is more important to have beauty in one's equations than to have them fit experiment. [. . .] It seems that if one is working from the point of view of getting beauty in one's equations, and if one has really a sound insight, one is on a sure line of progress."[15] As Richard H. Dalitz recalls, in Moscow in 1955, "When asked

13. Dirac discusses his use of the aesthetic properties of mathematical expressions as heuristic guides in Dirac (1982a). Krisch (1987), p. 51, reports: "Dirac stated that, '. . . the elegance of the formulation was very important in choosing the direction for one's research.' "
14. Dirac (1980a), p. 44. Chandrasekhar (1988), pp. 52–55, expresses doubts that in his search for a theory of gravitation Einstein was motivated by aesthetic factors to the extent to which Dirac supposes.
15. Dirac (1963), p. 47.

Two Challenges to Rationalism

to write briefly his philosophy of physics, he wrote on the blackboard 'PHYSICAL LAWS SHOULD HAVE MATHEMATICAL BEAUTY'."[16] It was at least in part on such a criterion that Dirac extended support to the theory of general relativity: "The foundations of the theory are, I believe, stronger than what one could get simply from the support of experimental evidence. The real foundations come from the great beauty of the theory. [. . .] It is the essential beauty of the theory which I feel is the real reason for believing in it."[17]

Thus, while logical positivists admit that aesthetic factors may play a part in the context of discovery but deny that they have any incidence in the context of justification, Dirac believes that the procedures typical of both stages make recourse to aesthetic considerations. If we wish to account for behavior such as that which Dirac notes, we will require a view of science more richly articulated than that of the logical positivists.

4. THE BOUNDARIES OF SCIENTIFIC BEHAVIOR

The second attempt that has been made by rationalists to dismiss the problem posed by scientists' aesthetic evaluation of theories is more subtle. Some authors admit that whereas aesthetic criteria are sometimes used by scientists in evaluating theories, this behavior is not scientific and thus does not enter the scope of descriptions of scientific practice.

Logical positivists defined scientific behavior so narrowly as to equate it with empiricist behavior. On their view, the task of scientists is to collect, process, summarize, and explain empirical data: all other actions are nonscientific and are induced by influences acting on science from without. For instance, Philipp Frank in the 1950s drew a distinction between two sets of criteria for theory evaluation, which he termed the "scientific" and the "extrascientific." The scientific criteria are agreement with observations and logical consistency: criteria of all other sorts are extrascientific.[18] On this view, any nonempirical concern that scientists may have is an external influence, perturbing science from its proper course. Since Frank would allocate aesthetic factors to the category of extrascientific criteria, he would maintain that they need not be considered in an account of scientific practice.

While few authors define scientific behavior as narrowly as the logical positivists, many continue to believe that whereas evaluating theories

16. Dalitz (1987), p. 20.
17. Dirac (1980b), p. 10.
18. Frank (1957), p. 359.

on the basis of empirical criteria pertains to science, appeals to aesthetic considerations do not. This belief is often expressed in the claim that scientists resort to aesthetic criteria only as tiebreakers, when they must choose among theories that empirical criteria have shown to be equally worthy. This claim is put forward by Fritz Rohrlich: "There is [. . .] great beauty in a physical theory. [. . .] It is that beauty which affects the credibility of one theory over another in the absence of more stringent criteria. For instance, the general theory of relativity is so beautiful that it is preferred over rival theories as long as those rival theories cannot account any better for the empirical facts."[19] This passage implies that aesthetic considerations would cease to carry weight if it were discovered that relativity theory accounts for the empirical facts any better or worse than its rivals. This view amounts to a denial of importance to aesthetic criteria: it allows them onto the scene only in cases where a scientist has ascertained, on empirical criteria, that they will have no consequence.

In reality, as we shall see, far from being wheeled up only when empirical criteria have shown the theories on offer to be equally worthy, aesthetic preferences often overrule the standard empirical criteria in scientists' choices among theories. The situation is therefore not that aesthetic criteria are applied once scientists have ascertained, on empirical standards for the acceptability of theories, which theories they may accept; rather, aesthetic and empirical criteria jointly determine scientists' standards for the acceptability of theories. Historical studies confirm that aesthetic considerations play a role in these decisions.[20]

The aesthetic factors of which we shall construct a model should therefore be considered as fully distinctive of science as scientists' logical or empirical concerns. This does not mean, of course, that no useful distinction can be drawn between scientists' empirical and aesthetic considerations; but it does mean that the distinctions we draw between them cannot be portrayed as a demarcation between the scientific and the extrascientific.

5. A Precursor: Hutcheson's Account of Beauty in Science

The reluctance of philosophers in the twentieth century to attribute roles of much importance to aesthetic judgments in scientific practice may be due partly to the lack of influential accounts of intellectual beauty in

---

19. Rohrlich (1987), pp. 13–14. For a similar opinion, see Osborne (1986a), p. 12.
20. For instance, Jacquette (1990) shows that Newton's view of what counts as a satisfactory law of nature was based partly on aesthetic considerations.

recent philosophy. Twentieth-century aesthetic theory, which has taken as its central concerns the beauty of artworks and of nature, has paid little attention to the beauty of intellectual constructs. Harold Osborne noted in 1964: "Nowadays the concept of intellectual beauty is not, I believe, commonly repudiated so much as neglected; few of the standard works on aesthetics pay more than lip-service to it and I know of none which has either attempted a deep analysis or given to it equal weight with sensory beauties in the framing of general aesthetic concepts."[21] However, the study of intellectual beauty has fallen into disregard only relatively recently: in eighteenth-century aesthetic theory, for instance, it held an important place. We will begin our investigation of scientists' aesthetic judgments by reviewing one of the most sophisticated eighteenth-century theorists of intellectual beauty, Francis Hutcheson. His views are relevant to our purposes since he explicitly extends his treatment to scientific theories, asserting that theories showing particular properties are to be regarded as beautiful.

Hutcheson's account of the beauty of intellectual constructs follows directly from his more general aesthetic theory. Hutcheson endorses an epistemological tenet that was popular in his time, that the qualities of objects are distinct from, and in fact the causes of, "ideas," which are the only immediate materials of sensory awareness. Beauty is such an idea, occasioned in the mind by particular qualities of external objects. As Hutcheson writes, "the word *beauty* is taken for *the idea raised in us,* and a *sense* of beauty for *our power of receiving this idea.*"[22] Hutcheson therefore understands "beauty" not as a property of objects but as the response of an observer's aesthetic perception to qualities of objects:

> Let it be observed that by absolute or original beauty is not understood any quality supposed to be in the object which should of itself be beautiful, without relation to any mind which perceives it. For beauty, like other names of sensible ideas, properly denotes the *perception* of some mind; so *cold, hot, sweet, bitter,* denote the sensations in our minds, to which perhaps there is no resemblance in the objects which excite these ideas in us, however we generally imagine otherwise.[23]

Having specified what kind of thing beauty is, Hutcheson turns to investigate which properties of objects cause the occurrence of ideas of beauty in the mind. "Since it is certain," he writes, "that we have *ideas* of

21. Osborne (1964), p. 160.
22. Hutcheson (1725), p. 34. For commentary, see Kivy (1976), pp. 57–60.
23. Hutcheson (1725), pp. 38–39.

beauty and harmony, let us examine what *quality* in objects excites these ideas, or is the occasion of them."[24] Hutcheson quickly reaches a conclusion: "The figures which excite in us the ideas of beauty seem to be those in which there is *uniformity amidst variety*. [. . .] What we call beautiful in objects, to speak in the mathematical style, seems to be in compound ratio of uniformity and variety: so that where the uniformity of bodies is equal, the beauty is as the variety; and where the variety is equal, the beauty is as the uniformity."[25] The property of "uniformity amidst variety" can be found in scenes in nature and works of art, but also in intellectual constructs: the latter are as capable of raising in us ideas of beauty as are concrete objects.

Hutcheson believes that in the practice of science we obtain special opportunities to perceive uniformity amidst variety and therefore to conceive ideas of beauty. The objects in which the scientist perceives uniformity amidst variety are located on three levels of increasing abstraction.

Objects on the lowest level are the entities and phenomena that constitute the subject matter of science. For instance, stars are arranged in the night sky with a high degree of uniformity amidst variety, and thereby give rise to ideas of beauty in observers. In order to derive a sense of beauty from these entities, ordinary observation of them is sufficient: no particular scientific theory or expertise is required, any more than it would be in order to come to see a landscape as beautiful.[26] Today the beauty of objects on Hutcheson's first level is recognized by astronomers who find beauty in views of celestial bodies and by chemists who speak of beautiful molecules.[27]

The objects on Hutcheson's second level of abstraction are natural regularities which are not directly to be seen in the phenomena but become apparent in the models or accounts put forward by theories. Although these regularities are endowed with uniformity amidst variety and can therefore raise in us ideas of beauty, they are apt to be perceived and therefore appreciated as beautiful only by observers who have some command of scientific theory. For instance, the astronomer sees into celestial motions more regularities than are apparent to the casual observer of the night sky. Isaac Newton's theory in celestial mechanics—which greatly impressed Hutcheson as well as most other eighteenth-century British empiricists—reveals regularities in the relations between such

24. Ibid., p. 39.
25. Ibid., p. 40.
26. Ibid., pp. 41–42.
27. Lynch and Edgerton (1988) discuss the aesthetic features of images of celestial bodies; Hoffmann (1990) surveys the properties of molecules that chemists regard as beautiful.

Two Challenges to Rationalism

quantities as the radii of the planets' orbits and the periods of their revolutions. While these regularities are properties of the phenomena, we are unable to perceive them but through the mediation of scientific theories. Writing that "these are the beauties which charm the astronomer, and make his tedious calculations pleasant,"[28] Hutcheson suggests that perception of this beauty is characteristic of technical work.

There is no doubt that beauty may be found in the regularities and other features that theories attribute to the world. For example, Charles Darwin's theory of evolution portrays a biological habitat as sustaining an intricate network of relations among organisms, which becomes visible through the theory's mediation; and it appears that Darwin felt aesthetic pleasure in viewing scenes in nature as such a network.[29] Similarly, geological theory may deepen our understanding and thus our aesthetic appreciation of landscapes.[30]

Properties of phenomena revealed by scientific theorizing have been cited as a source of aesthetic pleasure by artists through the centuries. For instance, much seventeenth-century English poetry regarded the universe as containing harmonies that are invisible to the uninformed eye but become apparent through the mediation of Aristotelian and Ptolemaic cosmology. In *Paradise Lost* (1667), John Milton appears to apprehend the heavenly motions in the light of astronomical models of them:

> That day, as other solemn days, they spent
> In song and dance about the sacred hill,
> Mystical dance, which yonder starry sphere
> Of planets and of fixed in all her wheels
> Resembles nearest, mazes intricate,
> Eccentric, intervolved, yet regular
> Then most, when most irregular they seem,
> And in their motions harmony divine
> So smooths her charming tones, that God's own ear
> Listens delighted.[31]

To the naked eye, celestial motions are haphazard; but through the intermediation of cosmological theory, they are revealed as exhibiting the

28. Hutcheson (1725), p. 43.

29. On Darwin's aesthetic appreciation of evolutionary phenomena, see Gruber (1978).

30. On the role of science in the aesthetic appreciation of landscapes, see Rolston (1995).

31. Milton, *Paradise Lost*, book V, lines 618–627. For further discussion of Milton's cosmological imagery, see Nicolson (1950), pp. 51–52, and (1956), pp. 80–109.

greatest regularity precisely when they appear to be most irregular. In the century after Milton, poetry came to behold the universe through Newtonian theory: both the view of the universe as a clockwork mechanism, inspired by the *Principia mathematica*, and the view of white light as a mixture of the spectral colors, presented by the *Opticks*, attracted literary responses.[32] In the twentieth century, poetry and other arts have similarly commented on, for instance, the world as portrayed by relativity theory.[33]

Lastly, Hutcheson believes that scientists perceive beauty in objects on a third level of abstraction: mathematical theorems and scientific theories themselves. He points out that some theorems and theories possess the property of uniformity amidst variety to an exemplary degree. He distinguishes general theorems and theories (which he calls "discoveries") from reports of individual observations, which might reveal truths but which show no unity: "Let us compare our satisfaction in such discoveries with the uneasy state of mind when we [. . .] are making experiments which we can reduce to no general canon, but are only heaping up a multitude of particular incoherent observations. Now each of these trials discovers a new truth, but with no pleasure or beauty, notwithstanding the variety, till we can discover some sort of unity or reduce them to some general canon."[34]

It is in theorems and theories of great generality that are known with certainty, or "universal truths demonstrated," that Hutcheson discerns the greatest capacity for aesthetic appeal: there is no other kind of entity "in which we shall see such an amazing variety with uniformity, and hence arises a very great pleasure." The reason for this is that in such constructs "we may find included, with the most exact agreement, an infinite multitude of particular truths, nay, often a multitude of infinities."[35] Hutcheson points out that this degree of beauty is found both in mathematics and the empirical sciences:

There is [. . .] beauty in propositions when one theorem contains a great multitude of corollaries easily deducible from it. [. . .] Such a theorem is the 35th of the 1st Book of Euclid, from which the whole art of measuring right-lined areas is deduced by resolution into triangles which are the halves of so many parallelograms [. . .]. In the search of nature there

32. On the inspiration of poetry by Newtonian theory, see Nicolson (1946), especially pp. 107–131, and Bush (1950), especially pp. 51–78.

33. Among the many studies of the impact of relativity theory on the arts, see Friedman and Donley (1985).

34. Hutcheson (1725), p. 49.

35. Ibid., p. 48.

Two Challenges to Rationalism

is the like beauty in the knowledge of some great principles or universal forces from which innumerable effects do flow. Such is *gravitation* in Sir Isaac Newton's scheme.[36]

Remaining faithful to the idea that the cause of ideas of beauty is the property of uniformity amidst variety, Hutcheson retraces the beauty of empirical theories to their generality and unifying power.

Hutcheson's treatment of the beauty of scientific theories offers answers to all the principal questions that we might think of posing. It specifies what sort of entity beauty is; it distinguishes between the beauty of a theory and the beauty of the phenomena that are the theory's subject matter; it describes the relation that a judgment that a theory is beautiful has to the properties of the theory; and it suggests which particular properties will lead scientists to regard a theory as beautiful. (As we shall see in Chapter 4, Hutcheson also presents a view of the relation between theories' empirical performance and scientists' aesthetic appraisals of them.) Although Hutcheson's discussion does not extend to social and historical aspects of scientific practice, his account entails claims about them too. For instance, it predicts that all scientists will recognize a given theory as beautiful to the same degree, provided only that they apprehend the theory correctly; and it entails that a judgment about the beauty of a given theory, if correctly passed, will never require revision. In the terms that we shall later use, Hutcheson claims that there is an aesthetic canon that all scientists in history share and on which all aesthetic evaluations of theories are and will be passed.

The interest of Hutcheson's account of the beauty of scientific theories is demonstrated by the fact that writers have continued to endorse or echo his suggestions. For instance, Adam Smith, who was a student of Hutcheson at Glasgow, suggested repeatedly in his writings that certain theories are beautiful in virtue of their unification of disparate observations, referring for instance to "the beauty of a systematical arrangement of different observations connected by a few common principles."[37] The significance of uniformity amidst variety in causing ideas of beauty has been affirmed by many mathematicians and scientists since then. Henri Poincaré, for example, asked, "What are the mathematical entities to which we attribute this character of beauty and elegance, which are capable of developing in us a kind of aesthetic emotion? Those whose elements are harmoniously arranged so that the mind can, without effort,

36. Ibid., p. 50. For further commentary on Hutcheson's treatment of the beauty of mathematical theorems and scientific theories, see Kivy (1976), pp. 97–99.
37. Smith (1776), pp. 768–769. For further discussion of the role that Smith attributed to aesthetic factors in scientific method, see H. F. Thomson (1965), pp. 219–221.

take in the whole without neglecting the details. This harmony is at once a satisfaction to our aesthetic requirements, and an assistance to the mind which it supports and guides."[38] Most recently, Nicholas Jardine has noted the capacity of scientific theories to "bring out" beauty in aspects of natural phenomena not apparent to the uninformed eye—objects on the second of Hutcheson's three levels of abstraction.[39]

I concur with parts of Hutcheson's account of the beauty of theories. For instance, I endorse his suggestion that observers attribute beauty to objects upon perceiving certain properties in them, and I accept his distinction between the aesthetic properties of theories and those of phenomena. But I find Hutcheson's account untenable for other reasons. I do not believe that, as he maintains, there is any one property that all scientists throughout history recognize as ensuring beauty in theories; I regard his account of the relation between scientists' aesthetic appreciations of theories and theories' empirical performance—a topic that we have yet to discuss—as inconsistent with evidence from scientific practice; and I draw conclusions more far-reaching than Hutcheson's about the role of aesthetic evaluations of theories in shaping the course of science.

38. Poincaré (1908), p. 59. For further discussion of Poincaré's views on the beauty of mathematical and scientific constructs, see Papert (1978), pp. 105–113.
39. Jardine (1991), pp. 209–212.

Two Challenges to Rationalism

Abstract Entities and
Aesthetic Evaluations

One of the older research students said the sweetest thing to me after my lecture, that he had never realized that there was anything aesthetic in Mathematics till one of my lectures. I was frightfully bucked.
                                                                    —Nevill Mott, *A Life in Science*

1. The Distinction between Theories and Their Representations

This book endorses what I have called the rationalist image of science. Models of the practice of theory assessment that are compatible with the rationalist image, such as the logico-empirical model of theory assessment that I discussed in the previous chapter, attach importance to a distinction between scientific theories and their representations. In this section and the two that follow, I outline this distinction and investigate some of its implications. This discussion will enable us to delineate more clearly the boundaries of our topic, scientists' aesthetic evaluations of theories.

Scientific theories are abstract entities. For this reason, although they are entities about which we can have knowledge, they are not entities that we can see or hear. To impart knowledge about a theory, we must first construct something that may be called a representation, rendering, or encoding of the theory in a certain language or code. This is the object that we read as a publication or hear as a lecture in order to acquire knowledge about the theory. Clearly, theories and representations of theories are entities of two different sorts: while theories are abstract entities, representations are concrete entities such as texts and utterances.

Scientific theories have properties. Among the possible properties of theories are those of being untrue, being complicated, and being probabilistic. Representations of scientific theories have properties too. Among the possible properties of representations of theories are those of being terse, of being in French, and of containing many diagrams. Since theories and representations of theories are entities of different sorts, no property of a theory can also be a property of a representation, even though some properties of theories may be given the same names as properties of representations. For instance, both a theory and a representation may be quantitative, but their properties remain distinct: a theory that is quantitative is so in virtue of making claims about the values of physical parameters, while a representation of a theory is quantitative if it contains mathematical equations. The distinctness of these properties is shown by the fact that a quantitative theory such as quantum theory can be given a purely qualitative representation.

According to rationalist models of theory evaluation, properties of theories' representations must not be held in the same regard as properties of theories. An evaluation of a theory should depend only on the properties of the theory itself, such as its degree of accord with empirical data and internal consistency, and not on properties of representations of the theory. For example, it would be unjustified for someone to hold against a theory any shortcomings of the lectures on it that he or she had heard. Of course, it is probably beyond our power to ensure that our opinions of theories are never influenced by properties of their representations; but to submit to such an influence is nonetheless, according to rationalist models of theory evaluation, unjustifiable.

The distinction between properties of theories and properties of theories' representations extends also to aesthetic properties. Representations of theories have many aesthetic properties. This is true most obviously of representations that take a pictorial form: Leonardo da Vinci's anatomical drawings encode and convey his theories in anatomy, and much nineteenth-century geological knowledge was contained in watercolors and etchings made by travelers.[1] The fact that such pictures have artistic qualities as well as containing sophisticated scientific claims makes it difficult to draw a sharp boundary between works of science and of art.[2] Even representations of theories in verbal form have notable aesthetic properties: historians and sociologists of science have become increasingly aware of the rhetorical and stylistic dimensions of scientific texts,

1. The techniques that Leonardo used to encode anatomical claims in visual form are reviewed by Veltman (1986), pp. 202–226; Renaissance anatomical illustration more broadly is studied by Ackerman (1985). On the geological content of pictures made by nineteenth-century travelers, see Stafford (1984), pp. 59–183.

2. That scientific and artistic representations of the world are not easily distinguishable is argued by Root-Bernstein (1984), Kemp (1990), and Edgerton (1991).

Abstract Entities and Aesthetic Evaluations

for example.[3] Nevertheless, rationalist models of theory evaluation prescribe that the aesthetic properties of theories' representations be given no weight in evaluating theories. That is why this book devotes no attention to such properties of representations of theories as the literary qualities of scientific texts. Whether rationalism can allow aesthetic properties of theories to be given a role in theory evaluation is, of course, the issue discussed throughout this book.

2. The Disregard of Abstract Entities by the Actor-Network Theory

Among the high-level models of scientific practice that are currently advanced as alternatives to the rationalist image, there are several that take a sociological or anthropological approach. Most of these models reject the distinction that I draw between theories and their representations. One of the most interesting models of this kind is the actor-network theory of scientific practice.[4]

According to the actor-network theory, all scientists as well as other human and non-human actors are linked in a network of causal interrelations. Scientific practice consists of attempts by groups of scientists to enroll other actors in the service of their interests, by producing and manipulating entities. The sole entities that can serve this purpose are concrete entities, since abstract or immaterial entities have no causal power—indeed, some proponents of the actor-network theory maintain that abstract entities do not exist. Among the concrete entities that are best suited to influence other scientists are inscriptions and texts, such as instrument readings and journal articles. In contrast, scientific theories, in virtue of being abstract entities, cannot influence the behavior of actors. These convictions explain how Bruno Latour and Steve Woolgar can interpret science largely as a system for producing inscriptions rather than theories, and why Latour thinks that what will shed most light on science is the analysis of particular scientific texts rather than of scientific theories.[5] Latour puts it as follows: "We do not think. We do not have ideas. Rather there is the action of *writing*, an action which involves working with *inscriptions* [. . .]; an action that is practiced through *talking*

3. Among the many recent studies of the rhetoric of scientific texts is Gross (1990).
4. As well as the actor-network theory, present-day sociological and anthropological models of science include the approaches known as sociology of scientific knowledge, ethnomethodology, reflexivity, and social epistemology. The debate among these approaches is carried forth in Pickering (1992).
5. Latour and Woolgar (1979), p. 88; Latour (1987), pp. 21–62.

to other people who likewise write, inscribe, talk [. . .]; an action that *convinces* or fails to convince with inscriptions which are made to speak, to write, and to be read."[6]

The actor-network theory offers a provocative and insightful view of science. However, I find it a less persuasive high-level model of scientific practice than the rationalist image, in part because of its insistence that only concrete entities such as inscriptions matter and that abstract entities such as scientific theories do not. It is impossible to account convincingly for the possession and propagation of knowledge without recognizing that items of knowledge are abstract entities that may be distinguished from the concrete entities in which they are expressed.

It is undeniable, for example, that some individuals and communities have the knowledge to produce at will certain physical occurrences, such as particular mechanical, electrical, and chemical effects. They can reproduce this knowledge in other individuals and communities by conveying certain entities to them. In which entities do they convey this knowledge? In the light of the statements that I cited above, I presume that adherents to the actor-network theory would say that this knowledge may be conveyed in a certain inscription or text. However, it could presumably be conveyed also in a faithful paraphrase or translation of that inscription. So it cannot be that the power of conveying the knowledge to produce a particular physical occurrence is exclusive to a given inscription: on the contrary, it must be that this power is common to all the inscriptions that are related to one another in a certain way. Having this power must be an abstract property of each of the inscriptions belonging to the group, since there is nothing concrete that is shared by them all. But this conclusion amounts to the claim that the knowledge to produce a particular physical occurrence is in fact conveyed in an abstract entity, of which each of the inscriptions belonging to the group is a distinct rendering. Nothing prevents us from identifying the abstract entity in question with a scientific theory, and the inscriptions with alternative representations of that theory.

By rejecting the concept of scientific theory, the actor-network theory loses the faculty of identifying a group of inscriptions as alternative representations of a particular theory; it is thus deprived of an explanation of the fact that the knowledge to produce a certain physical occurrence may be conveyed in any of the inscriptions belonging to a particular group. More broadly, by disregarding the concept of scientific theory in favor of that of inscription, adherents to the actor-network theory hamper the understanding of precisely those entities in which our knowledge

6. Latour (1984), p. 218.

Abstract Entities and Aesthetic Evaluations

of the physical world is contained. These shortcomings of the actor-network theory illustrate further the importance of distinguishing the properties of theories from the properties of the representations of theories, and of devoting due attention to the former.

3. Perceiving the Properties of Abstract Entities

The account of scientific theories that I have been sketching includes the following claims. Scientific theories are abstract entities, and abstract entities are not accessible to our senses of sight and hearing. Accordingly, we acquire information about theories not by seeing or hearing them, but by contemplating representations of them. These representations are accessible to our senses, since they are concrete entities such as publications and lectures. This account bequeaths the following puzzle, which has bothered all those, from Plato to the present, who have postulated abstract entities: in view of the fact that our senses have access only to concrete entities, how are we able to ascertain the properties of abstract entities?

Hutcheson faced this puzzle when he claimed that intellectual constructs such as mathematical theorems and scientific theories could have beauty. He attempted to solve it by postulating, alongside the "external senses" which he held responsible for the perception of external bodies, an "internal sense" which conceived ideas of beauty in the contemplation of abstract entities.[7] He cited evidence from musical aesthetics: "In music we seem universally to acknowledge something like a distinct sense from the external one of hearing, and call it a *good ear*."[8]

In present-day aesthetics, which does not favor the postulation of new sense organs, a different response to the puzzle has become standard. This relies on the notion of the transposition of properties from abstract entities to their representations. In many contexts, we apprehend certain properties in an abstract entity upon perceiving certain other properties in some concrete rendition of the entity. For instance, we see properties in a piece of music (an abstract entity to which we have no direct sensory access) upon perceiving other properties in one or more renditions or performances of it. For sure, some of the properties of a rendition originate with the rendition itself and cannot be retraced to the piece of music: a performance may for instance be mechanical or hurried. But other properties of the rendition allow us to apprehend properties of the piece

7. Hutcheson (1725), p. 34.
8. Ibid., p. 35. For commentary, see Kivy (1976), pp. 24–27.

of music itself: this is how we may come to see it as a fugue or as atonal, perhaps. Clearly, for this procedure to deliver accurate knowledge of the abstract entity, some properties of the rendition must stand in some specified relation to some properties of the abstract entity. Those who advance this response to the puzzle say that if we are to be able to apprehend properties of the abstract entity, they must be transposed into the rendition.[9]

On the model of this response, we could argue that we are able to apprehend some of the properties of a scientific theory upon perceiving some other properties in a representation of the theory in a concrete entity. Some of the properties of a representation of a theory in the form of a text are proper to the representation: the text might be in French, for instance. But some properties of a faithful representation will be owed to the mathematical structure, the logical parsimony, or other properties of the theory, that show through in the rendering. While our apprehension of these properties occurs via the rendering, they may legitimately be retraced to the theory.

A different possible response to the puzzle is the following. What we have been calling the representation of an abstract entity such as a scientific theory should be regarded not as a depiction of the entity but as an algorithm for creating a mental replica of it: consulting a representation of an abstract entity enables me to replicate it in my mind. A person's knowledge of the properties of an abstract entity is thus gained by examining a mental replica of it rather than the concrete representation. For an algorithm to yield a certain product, it is not required that the properties of the algorithm resemble those of the product: consequently, this response does not commit us to claiming that the properties of theories are transposed into concrete representations of them.

4. AESTHETIC VALUES, PROPERTIES, AND EVALUATIONS

In order to understand scientists' practice of evaluating scientific theories on aesthetic grounds, we need a working conception about what it is for an observer to pass an aesthetic judgment on an object. Analysis of such an act is the task of aesthetic theory. In this section, we draw from present-day aesthetic theory a conceptual apparatus for use in our investigation.

When we make an aesthetic appreciation of an object, we refer to enti-

9. The transposition of aesthetic properties is discussed by Wollheim (1968), pp. 74–84.

ties of the following two kinds, among others: perceptible properties and aesthetic values. Our aesthetic appreciation must refer to properties of the object if it is to be an appreciation of that object, and it must refer to values if it is to be evaluative. To specify what these properties and values to which we appeal are, we must provide answers to questions such as the following:

1. Is beauty a property of objects, a value, or an entity of some other kind?

2. Does aesthetic value reside in objects of perception, or is it projected into them by observers?

3. What is the relation between aesthetic value and the intrinsic properties of objects to which an aesthetic appreciation refers?

4. How can aesthetic appreciations of objects by different observers show the great diversity that is characteristic of aesthetic discussions, when they all refer to properties of objects that presumably are independent of the identity of observers?

The answers that I give for our purposes are as follows.

1. Although in some traditions, such as Platonism, beauty is understood as a property that is intrinsic to some objects, I regard it as an aesthetic value. Values are respects in which things can be good, important, or desirable. If beauty were a property of objects, the statement that a given object is beautiful would be a purely descriptive report, on a par with the statement that a given object is spherical. If beauty is a value, the statement that a given object is beautiful has an evaluative component, implying judgments about the object's goodness, importance, or desirability.

Incidentally, beauty is not the sole aesthetic value that we may conceive: another value that features in the appraisal of artworks is artistic merit. The difference between beauty and artistic merit is shown by the following considerations. An artwork might not be beautiful, but might yet possess artistic merit, in virtue of having, say, great originality, which does not of itself confer beauty. On the other hand, if an object is perceptually indistinguishable from a beautiful object, it has to be deemed beautiful too; but it may be that only one of these two objects has artistic merit, if, for example, one has had a great influence on the development of art while the other is merely a later replica. This suggests that whereas a judgment of beauty is based entirely on properties that are perceptible in the object under evaluation, a judgment of artistic merit may refer to relational properties that are not manifest in the object, such as the property of having had a particular history or standing in particular relations to other artworks.

It might be possible to identify an aesthetic value that is the counterpart for scientific theories of the artistic merit of artworks. By analogy with artistic merit, this aesthetic value would be acquired by a scientific theory in virtue of possessing certain relational properties that may not be manifest in it, such as the property of having had a particular influence on the development of science. However, the sole aesthetic value that this book will discuss is beauty.

2. The question of where value is located alludes to a controversy in aesthetics and ethics between two doctrines. One doctrine, objectivism, contends that value resides in the world and is available to be encountered by observers. The alternative, projectivism, claims that value is not to be found in the world but is instead projected into it by observers, as a reflection of their responses (such as judgments or emotions) to objects.[10]

Projectivism is defended more commonly in ethics than in aesthetics. In his defense of ethical projectivism, John L. Mackie holds that something is objective only if it is fully describable in terms of properties that can be understood without reference to their effects on sentient beings. Since, according to Mackie, values cannot be described fully without referring to certain properties' effects on sentient beings, he concludes that values are not objective. As evidence for his view, Mackie cites the diversity of the value judgments passed by different observers about given objects. He suggests that this diversity is better explained by supposing that it reflects differences in values held to by the observers, rather than merely the observers' differing responses to values that are located in the world.[11]

In this book, I hold to a projectivism about aesthetic value analogous to that of Mackie in ethics: I presume that the value to which the aesthetic appreciation of scientific theories refers does not reside in the theories themselves but rather is projected into theories by individual scientists, scientific communities, and observers of science. This amounts to the claim, which I think is very plausible, that we cannot fully describe a scientific theory's aesthetic value without referring to the effect of properties of that theory on scientists or other observers. My grounds for espousing projectivism are analogous to those of Mackie. I discern much diversity in the aesthetic responses of scientists and others to theories. Objectivism would involve explaining this diversity entirely as the effect of differences in scientists' reactions to aesthetic value located intrinsically in scientific theories. A better explanation of this diversity, I feel, is

10. Some alternatives to objectivism and projectivism about aesthetic value are explored in Wollheim (1968), pp. 231–240.

11. Mackie (1977), pp. 15–49. Projectivism for moral values is defended also by Blackburn (1984), pp. 181–223. Projectivism in aesthetics is explored by McDowell (1983).

Abstract Entities and Aesthetic Evaluations

to say that aesthetic value is projected into theories in differing amounts or intensities by different scientists and scientific communities. Thus, the amount or intensity of aesthetic value that a given theory comes to have for different observers may vary.

Of course, the impression is easily formed that values are objective. Indeed, the semantics of "beauty" and its derivatives is mostly objectivist: we speak of entities as "having beauty" or "being beautiful," for example. But the fact that many observers have the impression that values are objective does not discredit projectivism about aesthetic or moral values. Simply, the phenomenology of human perception may be such that most observers, in projecting aesthetic or moral values into an object, feel themselves to be encountering a value located in the object.[12] Although I will continue to use "beauty" and related words in the standard objectivist locutions, I regard these as no more than abbreviations for more accurate, strictly projectivist descriptions of the passing of value judgments.

3. What is the relation between the aesthetic value of beauty and the intrinsic properties of objects? Whether, presented with a particular object, we project into it aesthetic value must depend in part on the object's intrinsic properties. However, it is likely that the aesthetic response that an object prompts in an observer is evoked by only some of the object's properties. I name "aesthetic properties" those properties intrinsic to an object that evoke the aesthetic response of observers to that object, and, more specifically, contribute to determine whether observers project beauty into that object. For example, a painting has many intrinsic properties, some of which—properties relating to its composition, draftmanship, or coloring—will contribute to determine our aesthetic response. In my usage, these are the painting's aesthetic properties. So whereas an object's aesthetic properties are properties of that object intrinsically, in the sense of belonging objectively to it, these properties are aesthetic only in virtue of evoking the aesthetic response that observers have to the object.

Notice that, for a property to count on this definition as aesthetic, it is not required that it should be pleasing. Aesthetic responses include not only sensations of pleasure and projections of aesthetic value into objects but also sensations of displeasure and denials of aesthetic value. A property that evokes any of these responses in an observer counts on my definition as aesthetic. Thus, the fact that an object has aesthetic properties does not ensure that a given observer will find it beautiful: an ob-

12. Blackburn (1985) argues that projectivism in ethics is able to account for the objectivist "feel" of moral judgments.

server will regard an object as beautiful only upon discerning in it specific aesthetic properties, viz., those that he or she values positively.

My definition of aesthetic property as any of the intrinsic properties of objects that evoke an aesthetic response in observers would be rejected by many other writers. The view that has long prevailed in aesthetics is that the aesthetic properties of objects are qualities, such as "grace," "daring," and "mournfulness," that are pleasing or displeasing intrinsically and that are attributed to particular objects by discerning beholders.[13] Adherents of this view enter into elaborate discussions of how an object's intrinsic properties, which they regard as nonaesthetic, can support a beholder's attribution to it of aesthetic properties. My disagreement with this view centers on the question what the inherent attributes of aesthetic properties are. On the prevailing view, aesthetic properties have an intrinsic evaluative dimension, since they are intrinsically pleasing or displeasing, but they are not intrinsic properties of objects of perception, amounting rather to interpretations of objects by beholders. By contrast, the properties that I recognize as aesthetic are properties of objects intrinsically, and are evaluative only in the sense that they evoke aesthetic evaluations of objects in observers. On my view, such terms as "grace," "daring," and "mournfulness" are not names of properties of objects, but elliptical characterizations of a beholder's response to an object's aesthetic properties.

4. Lastly, how are we to explain the diversity of observers' aesthetic responses to objects? I address this question in the specific case of the aesthetic perception of scientific theories. To explain the diversity of scientists' aesthetic responses to theories, we must consider how a scientist is moved to project beauty into a particular theory. I suggested in my answer to question 3 that scientists pass aesthetic judgments on theories in response to the properties that they perceive in the theories. Upon perceiving particular properties in a theory, a scientist projects beauty into it. The question then remains, what ensures that a scientist picks out certain properties as those that warrant the projection of beauty into theories?

The fact that a scientist picks out particular properties of theories as warranting projections of beauty will in this book be taken as a consequence of the scientist's holding to particular aesthetic criteria, which refer to the properties in question. Simply, different scientists or scientific communities hold to different criteria on the basis of which they pass aesthetic judgments on theories and, in particular, decide on the amounts

13. Those who take the view of aesthetic properties described here include Sibley (1959), Hungerland (1968), Beardsley (1973), and Goldman (1990).

Abstract Entities and Aesthetic Evaluations

or intensities of the value of beauty that they will project into given theories.

The nature, origin, and mode of development of these aesthetic criteria will be elucidated progressively in this book. Let us summarize the tenets of the aesthetic theory outlined so far. Aesthetic values, such as beauty, are not located in the world, but rather are projected into objects by observers. An object of perception, such as a scientific theory, may have, among its intrinsic properties, some which evoke aesthetic responses in observers, for example inducing them to project the value of beauty into the object. I deem such properties to be the object's aesthetic properties. A scientist is moved to project beauty into a theory by virtue of holding to one or more aesthetic criteria, which attach aesthetic value to properties that the theory has. Finally, I explain the diversity of scientists' aesthetic responses to scientific theories on the assumption that different scientists hold to different sets of such aesthetic criteria.

5. Aesthetic Criteria and Canons

I have said in the previous section that I interpret a scientist's being moved to project beauty into a theory as a consequence of his or her holding to one or more aesthetic criteria, which attach aesthetic value to properties of the theory. For every property that a theory might show, we can envisage a corresponding aesthetic criterion. The criterion for a property P might take the form "If a theory has P, attach more aesthetic value to it than if, other circumstances being equal, it did not." Such criteria offer evaluations of theories: they may be applied in cases of theory assessment and may contribute to deciding cases of theory choice.

We may assume a scientist to hold to as many aesthetic criteria of this form as there are properties of theories to which he or she makes an aesthetic response. The aesthetic criteria to which a scientist holds constitute what I shall call his or her aesthetic canon. A scientific community too can be described as possessing such a canon, if there is sufficient unity among the aesthetic canons of its members to make this attribution meaningful. Aesthetic canons, like individual aesthetic criteria, may be used to appraise theories and to choose between them.

The aesthetic criteria that make up a canon can be assumed to possess weightings that differ from one another. In other words, whereas each criterion attaches aesthetic value to a particular property of theories, one criterion might weigh or be worth more than another within the canon. Thus, suppose that a scientist must choose on aesthetic grounds between two theories that are equally attractive, except that one theory exhibits the property P while the other exhibits in its place the property Q. If both these properties are valued by the scientist's aesthetic canon, and the criteria attaching value to them have equal influence, the scientist will regard the two theories as equally worthy. But if these criteria have different weightings, the criterion with the greater weighting will prevail, and the scientist will prefer the theory that exhibits the property to which this criterion refers.

A fully expressed aesthetic criterion will therefore both refer to a possible property of theories and carry a certain weighting, which determines the influence of the criterion in theory choice. A criterion can thus be represented by a pair of items of information: the specification of a possible property of theories such as P, and the specification of a weighting W_P. A canon, which is a set of such criteria, would consequently take the following form:

$$P, W_P$$
$$Q, W_Q$$
$$R, W_R$$

.

.

.

We have spoken so far as though a given scientist's aesthetic canon contained a small number of criteria, corresponding to the few properties of theories to which he or she attaches aesthetic value, and of course this is a natural way of imagining it. But a useful generalized way of depicting scientists' aesthetic canons is to regard them as composed of a very large or even infinite number of entries: one for each of the properties of scientific theories to which aesthetic value could conceivably be attributed. In the case of any given scientist, the overwhelming majority of these criteria will carry a weighting of zero, since scientists typically attach aesthetic value to only a few of the conceivable properties of theories and are indifferent to the rest. The advantage of this depiction of aesthetic canons is that any evolution of a canon can thereby be represented as only a change in the weightings attached to each of the criteria in the very long or infinite list.

6. IDENTIFYING WHICH PROPERTIES OF THEORIES ARE AESTHETIC

While I have indicated that I define an aesthetic property as one that evokes aesthetic responses in observers, I have not yet specified how I propose in practical cases to recognize which properties of scientific theories are on this definition aesthetic. I shall rely on two criteria.

First, I shall judge a property of a theory to be an aesthetic property if scientists in the relevant disciplines react to it publicly as aesthetic, for example by declaring that they attach aesthetic value to it, by citing it in an act of theory evaluation that they describe as aesthetic, or by applying to it standard terms of aesthetic appreciation, such as "beautiful," "elegant," "pleasing," or "ugly." I regard these acts as amounting to aesthetic responses to the property in question, so any property of theories that prompts these acts in scientists satisfies in a straightforward way my definition of aesthetic property. Of course, in many cases a scientist will express aesthetic pleasure or displeasure not at specified properties of theories, but at a theory or group of theories in its entirety, speaking for example of "a beautiful theory." In such cases we may have to infer from circumstantial evidence which properties of the theories in question have induced this aesthetic response.

My willingness to rely on the acts of scientists to indicate which properties of theories are aesthetic might invite two objections. The first is that a scientist's application of aesthetic terms to a property of theories may not in fact amount to an aesthetic response to that property; the second is that a scientist's public acts and statements may misrepresent aesthetic preferences among properties of theories that he or she holds privately. Let us consider these objections in greater detail.

The first objection is supported by the observation that people sometimes use aesthetic terms to express nonaesthetic judgments of entities. Among scientists, aesthetic terms are sometimes used in appraisals of such things as the accuracy and persuasiveness of empirical data. For example, in his laboratory notebooks, Robert A. Millikan marked some of the data from his experiments on the electron charge with such words as "Beauty" and "Beautiful."[14] No one would conclude from this that Millikan had noticed a particular elegance in the figures at which he had arrived; rather, he was expressing the sentiment "This is just the sort of data that I had hoped to obtain." The objection, in short, is that people experience satisfaction at any outcome that promotes their interests, and sometimes they express this emotion in a pseudo-aesthetic vocabulary.

I acknowledge the legitimacy of this objection: it is likely that some occurrences of aesthetic terms in scientists' remarks about theories express not aesthetic appraisals of the theories but judgments of other sorts. Because of this fact, not all occurrences of aesthetic terms in scientists'

14. These remarks of Millikan are reproduced in Holton (1978), p. 64, and discussed on pp. 25–83.

remarks are pertinent to our discussion. I believe, however, that very many such occurrences, and specifically those that I cite in this book, embody appraisals of theories that are genuinely aesthetic. After all, the conceptual entities that we call scientific theories are available to aesthetic evaluation on a par with all other entities, so it should be expected that, on some occasions, they will in fact be subjected to aesthetic evaluation. On a principle of charity of interpretation, we should read scientists' remarks that purport to express an aesthetic evaluation of a theory as genuinely doing so, except where textual evidence or testimony suggests the contrary.

The second objection mentioned above consists in the claim that a scientist's publicly attributing aesthetic value to a particular property of theories, or endorsing a particular theory on aesthetic grounds, may misrepresent aesthetic preferences that he or she holds privately. This claim might be supported by the general observation that scientists' public acts and pronouncements are sometimes at variance with their private convictions. This objection fails to disqualify the use of scientists' public acts and statements as evidence in our investigation, however. It fails because our aim is to construct a model of the aesthetic evaluations of theories that have affected the course of science, and particularly the succession of theories that have been adopted in scientific communities. In order to have affected science thus, an evaluation of a theory must have been given public expression in acts or statements. It is therefore precisely the aesthetic responses to theories that scientists have made publicly that we must study and account for. The further task of discovering whether a particular scientist's public expressions departed from his or her privately held aesthetic preferences I leave to other investigators: the scientist's intellectual biographers.

On these grounds, in spite of the two objections raised, I consider it valid to make cautious appeal to scientists' public acts and statements in ascertaining which properties of theories evoke aesthetic responses in observers, and therefore which properties of theories are aesthetic on my definition.

The second criterion by which I propose to recognize properties of theories as aesthetic is a little broader: a property is aesthetic if, in virtue of possessing that property, a scientific theory is liable to strike beholders as having a high degree of aptness. The justification of this criterion is that, in many philosophies of art, the beauty of an object is explicated as its aptness or the aptness of its elements. Aptness has been central to notions of beauty since classical times: Greek art theorists, including Plato, knew it as *prepon*, and Roman writers, such as Marcus Vitruvius, as *decor*.[15] It is because of the importance given to these concepts in classical times that, for instance, the consistent use of architectural orders was regarded as imparting beauty to buildings. Hutcheson's suggestion that what we find beautiful is uniformity amidst variety may also be interpreted as alluding to aptness in objects of perception. Even today, to explain on what grounds one regards an object as beautiful, it is common to refer to it or its elements as notably appropriate, fitting, proper, or seemly.

This connection between beauty and aptness may be taken to hold also in scientific theories. Indeed, some scientists, including Werner Heisenberg, define beauty in theories as "the proper conformity of the parts to one another and to the whole."[16] Many other scientists attribute the aesthetic pleasure that they receive from theories to what they call an aptness of the theories. Some scientists perceive in certain theories an aptness so intense that they describe it as perfection or inevitability, as Steven Weinberg illustrates:

> In listening to a piece of music or hearing a sonnet one sometimes feels an intense aesthetic pleasure at the sense that nothing in the work could be changed, that there is not one note or one word that you would want to have different. In Raphael's *Holy Family* the placement of every figure on the canvas is perfect. This may not be of all paintings in the world your favorite, but, as you look at that painting, there is nothing that you would want Raphael to have done differently. The same is partly true (it is never more than partly true) of general relativity. Once you know the general physical principles adopted by Einstein, you understand that there is no other significantly different theory of gravitation to which Einstein could have been led. [. . .] The same sense of inevitability can be found (again, only in part) in our modern standard model of the strong and electroweak forces that act on elementary particles.[17]

Given the connection between beauty and aptness, it is justified to regard properties by virtue of which theories are liable to appear apt as aesthetic properties. This second criterion for recognizing aesthetic properties will lead us in the next chapter to treat as aesthetic some properties of theories, such as their metaphysical allegiances, that are less commonly considered aesthetic but nevertheless share many of the features of properties of theories that are unquestionably aesthetic.

15. See Pollitt (1974), pp. 217–218 on *prepon*, and pp. 341–347 on *decor*.
16. Heisenberg (1970), p. 174; Lipscomb (1982), p. 4; Chandrasekhar (1987), p. 70.
17. Weinberg (1993), pp. 107–108.

The Aesthetic Properties
of Scientific Theories

1. CLASSES OF AESTHETIC PROPERTIES

While any of the properties of scientific theories may evoke aesthetic responses in scientists, in practice relatively few properties do so. In accord with the definition that I gave in the previous chapter, I call these the aesthetic properties of theories. In this chapter, we survey those that have had the greatest influence on scientists in history.

The aesthetic properties of theories are grouped in what I shall call classes: for example, all the forms of symmetry that theories may show fall into a class. Classes of aesthetic properties constitute the headings under which scientists pass aesthetic evaluations of theories. It is not, however, in virtue of perceiving a class of aesthetic properties that a scientist passes an aesthetic evaluation of a theory: such an evaluation is evoked by and grounded in the actual aesthetic properties that the theory shows. For example, a scientist might compare two theories under the heading of the class of symmetry properties, but would prefer one for the particular form of symmetry that it shows. In order to predict how a scientist will choose between competing theories, it is not enough to be told that the scientist's aesthetic canon makes reference to symmetry: after all, virtually any theory could be said to have some symmetry or other. One needs to be told which form or forms of symmetry are held up by the scientist's canon as desirable.

The distinction between aesthetic properties and their classes should help to allay the skepticism of those who claim that it is impossible for scientists' aesthetic preferences to play a role in theory assessment, since they are never defined sharply enough.[1] This skepticism might be justified if scientists' preferences referred only to classes of aesthetic properties, such as "symmetry"; but as we shall see, scientists' preferences are specified far more precisely than this.

The grouping of aesthetic properties into classes manifests itself also for entities other than scientific theories. For instance, the novelist E. M. Forster wrote once that he valued novels whose plots show symmetry. Symmetry here is a class of aesthetic properties rather than a property that can be said to be present in certain novels and absent in others. Before we can predict precisely which novels Forster would admire for their symmetry, we must know which member of this class of properties he finds desirable. He might value the novels of Thomas Hardy because in them the landscape has a mood that mirrors that of the protagonists, or those of Charles Dickens because in them events are recounted from the viewpoint of more than one participant. In fact, it emerges that Forster admires *Thaïs* by Anatole France and *The Ambassadors* by Henry James because in these novels two protagonists come gradually to exchange psychological stances, so that the plot assumes "the shape of an hour-glass."[2]

This chapter discusses four classes of aesthetic properties of theories: form of symmetry, invocation of a model, visualizability/abstractness, and metaphysical allegiance. Discussion of a fifth class, form of simplicity, is postponed to Chapter 7, so that it may benefit from ideas to be introduced in intervening chapters. For each of these classes, we shall examine several properties to which scientists have attached aesthetic value and several examples of aesthetic evaluations of theories based on these properties.[3]

I do not claim that this list exhausts the classes of properties to which scientists refer when they pass aesthetic evaluations of theories. Neither do I claim that every aesthetic property can be allocated uniquely to one of the five classes that I name. For instance, the discussions among sixteenth- and seventeenth-century astronomers about the harmony of their theories could be interpreted as pertaining primarily to the theories' simplicity properties, their symmetries, or their metaphysical allegiances.

It is my contention in this chapter that properties belonging to each of

---

1. For an expression of the belief that scientists' aesthetic preferences are never defined sharply enough to be used in theory assessment, see Laudan (1984), p. 52; for a response, see Martin (1989).

2. Forster (1927), pp. 102–105.

3. I have previously surveyed the aesthetic criteria used by scientists in theory evaluation in McAllister (1989), pp. 30–36; other reviews of such criteria are in Osborne (1984) and Engler (1990), pp. 28–31.

these classes have profoundly influenced theory choice in the sciences since at least the Renaissance. By establishing this, I aim to demonstrate that the evaluation of theories on aesthetic grounds is an influential component of scientific practice and that any account of science that fails to refer to it must be considered incomplete.

2. FORM OF SYMMETRY

A structure is symmetric under a certain transformation just if that transformation leaves the structure unchanged.[4] Symmetries are displayed by many physical objects and natural phenomena. Some objects, such as molecules, snowflakes, and galaxies, have symmetries in virtue of being invariant under rotations or reflections. The symmetries associated with particular transformations in physics, such as the Galilean or the Lorentz transformations, are examples of symmetries of phenomena. When physicists speak of a theory's being invariant under the Lorentz transformations, for instance, what they mean is that the theory attributes this invariance to particular phenomena.[5]

Symmetries can be displayed also by abstract objects such as intellectual creations. For example, a fugue—a musical composition in which three or more voices carry phrases that are variants of the same theme—has several approximate symmetries: interchanging the phrases among the voices leaves the composition roughly unaltered.

The symmetry properties that are of interest to our investigation are, of course, properties of the abstract objects called scientific theories. A scientific theory can be said to have a symmetry if applying a transformation to the theory's conceptual components (its concepts, postulates, arguments, equations, or other elements) leaves the theory's content or claims unaltered. For the most part, we shall be talking of symmetries that are approximate rather than perfect, in virtue of which a transformation leaves a theory approximately unaltered. The symmetry properties of theories are distinct from the symmetry properties of phenomena: the former can provide grounds for choosing between two theories that describe the same phenomenon.

Here are four examples of the symmetries that scientific theories can have. First, a near-perfect symmetry is shown by the set of equations in classical electrodynamics known as Maxwell's equations, formulated by

---

4. The classic treatment of symmetry is Weyl (1952).
5. Symmetries of phenomena come under philosophical attention in, for example, Falkenburg (1988) and van Fraassen (1989), pp. 233–289.

James Clerk Maxwell and others. In the case of zero electric sources, Maxwell's equations are:

$$\text{curlE} + 1/c\,\mathbf{\dot{H}} = 0 \qquad \text{curlH} - 1/c\,\mathbf{\dot{E}} = 0$$
$$\text{divH} = 0 \qquad\qquad \text{divE} = 0$$

where **E** is the electric field vector, **H** the magnetic field vector, c the velocity of light, and curl and div are vector operators. Interchanging **E** and **H** in these equations leaves their content nearly unaffected.[6] Many physicists regard this symmetry as one of the aesthetic virtues of Maxwell's equations.[7] It seems likely that Maxwell himself was favorably struck by it. He had an aesthetic appreciation for mathematical structures: "I always regarded mathematics as the method of obtaining the *best* shapes and dimensions of things; and this meant not only the most useful and economical, but chiefly the most harmonious and the most beautiful."[8] In this light, it is reasonable to endorse the conclusion of Roger Penrose: "It would seem that the symmetry of these equations and the aesthetic appeal that this symmetry generated must have played an important rôle for Maxwell in his completion of these equations."[9]

Second, at the heart of the arguments on which Albert Einstein rejected certain features of classical physics and justified relativity theory lay considerations of the forms of symmetry that he felt it was appropriate to demand of physical theories—a more detailed discussion of this will come in Chapter 11. Symmetry arguments were used also by Hermann Minkowski in further developing relativity theory, though Minkowski's arguments hinged on geometrical considerations rather than on physical considerations of the sort to which Einstein appealed.[10]

My third example is the symmetry embodied in wave–particle dualism. In 1900, in order to account for the spectrum of black-body radiation, Max Planck suggested that electromagnetic radiation such as light, which was then viewed exclusively as a wave phenomenon, exhibits some particle properties, in that its energy varies by discrete amounts or quanta. Planck asserted that the value of the quantum of radiant energy E is proportional to the frequency of the radiation f:

$$E = hf$$

6. The symmetry of Maxwell's equations is further discussed in Rosen (1975), pp. 101–102.

7. For a physicist's aesthetic appreciation of the symmetry of Maxwell's equations, see Tsilikis (1959), pp. 92–94.

8. Letter to Galton, quoted in Hilts (1975), p. 59.

9. Penrose (1974), p. 271.

10. On Minkowski's aesthetic preferences, see Galison (1979), pp. 103–105.

where h is the Planck constant. This equation associates a particle property, the discrete energy, with a wave property, the frequency. In 1923, Louis de Broglie suggested that, correspondingly, particles possess wave properties, in virtue of which they show diffraction and interference. De Broglie asserted that a particle that has momentum mv may be described as a wave of a wavelength λ given by the following formula:

$$\lambda = h/mv$$

This formula associates a wave property, the wavelength, with a particle property, the momentum, mirroring Planck's equation. De Broglie came to his suggestion chiefly on the basis of symmetry considerations, or "purely on grounds of intellectual beauty," in the words of Michael Polanyi.[11] Empirical support for de Broglie's theory remained scarce until 1927, when the idea that particles show wave properties was exploited in wave mechanics. In the intervening years, the physics community's favorable attitude to de Broglie's theory was based mainly on the observation that it imparted a striking symmetry to physical theory.

Fourth, Harold G. Cassidy speaks of the satisfaction that he derived from perceiving symmetry in a theory of electron exchange polymers that he developed: "While I was listening to a piano concerto, the idea suddenly occurred to me that it should be possible to prepare electron exchange polymers. I was at once certain that this was feasible, and I felt the fitness of the idea in complementing the already well-known proton exchange polymers. [. . .] Once the symmetry of the relationship became apparent to me I experienced great pleasure and excitement."[12] Cassidy's account of this episode suggests that his perception of the symmetry acted not only as a stimulus to investigate the idea further but also as grounds for giving a favorable evaluation of the theory.

These examples show that scientific theories can exhibit widely differing forms of symmetry. The symmetry of Maxwell's theory consists in the making of similar claims about distinct physical parameters. The symmetry that Einstein valued, and which he judged classical physical theory to possess to an insufficient degree, is one in virtue of which a theory offers explanations of the same form for events deemed physically equivalent. The form of symmetry exhibited by de Broglie's theory (as well as, I think, by Cassidy's) is one in virtue of which a theory, if it attributes properties previously associated with one entity to a second

11. Polanyi (1958), p. 148. Evidence about the role of symmetry considerations in de Broglie's reasoning is given in Mehra and Rechenberg (1982–1987), 1:586–587.
12. Cassidy (1962), p. 57.

The Aesthetic Properties of Scientific Theories

entity, also attributes the corresponding properties of the latter entity to the former. The form of symmetry valued by Einstein is identical neither to the form exhibited by Maxwell's equations nor to the form shown by de Broglie's theory.

The forms of symmetry that theories exhibit should be counted among their aesthetic properties. Physical scientists frequently cite symmetry properties as grounds for regarding theories as beautiful. As the physicist Anthony Zee writes, "Given two theories, physicists feel that the more symmetrical one, generally, is the more beautiful."[13] More broadly, the symmetry properties of theories are liable to raise in beholders a sense of aptness. In fact, symmetries have been regarded as aesthetic properties of entities since ancient Greek times.[14]

3. Invocation of a Model

When we say that a scientific theory invokes a model, we mean that the theory implicitly or explicitly posits an analogy between the domain of phenomena that it is attempting to describe or explain and a certain other domain of phenomena, typically one that is better understood or more familiar. This more familiar domain of phenomena is called the model of the theory or the source domain of the analogy, while the domain of phenomena to which the model is applied is the analogy's target domain.[15]

According to the structure-mapping theory of analogy developed by Dedre Gentner, analogies function as follows. Any domain of entities exhibits an ordered series of relations: properties (which can be considered as relations of order zero), first-order relations holding between properties, second-order relations holding between first-order relations, and so on. An analogy maps entities of the source domain onto entities of the target domain. Furthermore, it maps onto the target domain a certain number of the relations exhibited by the source domain. On Gentner's account, the higher the order of the relations that the model preserves (even at the expense of lower-order relations), the better the model. Thus, properties of entities are the elements that are least important for the model to preserve.[16]

---

13. Zee (1986), p. 13. Some of the aesthetic aspects of symmetry considerations in theory appraisal are further discussed in Rosen (1975), pp. 120–122.

14. For a discussion of symmetry in general aesthetics, see Osborne (1986b). On notions of symmetry in ancient Greek theory of art, see Pollitt (1974), pp. 14–22.

15. The literature on models in science is reviewed by Leary (1990a).

16. Gentner (1983); Gentner and Jeziorski (1989).

As an illustration, Gentner cites the solar system analogy posited by Rutherford's theory of the atom. Entities of the analogy's source domain, the solar system, are mapped onto the entities of the target domain, the atom: the sun is mapped onto the atomic nucleus, and the planets are mapped onto electrons. The analogy, while disregarding many properties of entities of the source domain, preserves some higher-order relations of it. For instance, the relation "The sun's attraction of the planets causes them to orbit it," which holds in the source domain, is mapped onto the relation "The nucleus's attraction of the electrons causes them to orbit it," holding in the target domain.

Whether a particular scientific theory invokes a model, and which kind of model it invokes, are two of the factors that affect scientists' evaluations of theories. Many theories have gathered endorsement partly in virtue of the fact that they invoke a particular model; many other theories have repelled scientists by proving unsusceptible to being treated in terms of cherished models.

An illustration of scientists' preferences in the matter of the models that theories invoke is offered by early-nineteenth-century accounts of heat. In the 1820s, Pierre-Simon de Laplace and Joseph Fourier each developed a mathematical theory of heat. They worked in the same tradition, seeking to mathematize phenomena in order to apply to them the techniques that Newton had exploited with much success in celestial mechanics. But the requirements that Laplace and Fourier imposed upon physical theories were otherwise somewhat different. For Laplace, a theory could not be considered acceptable unless it offered a model of phenomena: his caloric theory presented a model of heat as a fluid. By contrast, Fourier denied that it was important for theories to offer models of the phenomena that they described: he preferred a purely analytical approach. He maintained that, as long as a theory was well confirmed by empirical data, it was sufficient for it to put forward what he called "the equations of the phenomena" without attempting to interpret them by analogy with some other domain of physics.[17]

There is clearly a difference between prescribing that theories should invoke some model or other and insisting that theories should be interpretable in terms of a model of a particular kind, such as a mechanistic, electrical, or biological one. After all, it is always possible to find a theory that stands in a relation of analogy to a given theory. In fact, most scientists who voice feelings in this matter express a preference for theories that invoke models of a particular kind. For them, a theory that invokes

17. On model building and positivist attitudes in nineteenth-century theories of heat, see Kargon (1969), pp. 424–430, and Bellone (1973), pp. 29–53.

The Aesthetic Properties of Scientific Theories

the wrong kind of model has no more merit than one that invokes no model. In many cases, the scientists who prescribe that theories should invoke models of a particular kind are motivated by a conviction that one of the sciences is fundamental and should provide the explanatory principles of all other sciences. For example, the style of theorizing in nineteenth-century physics known as mechanicism prescribed that all phenomena, such as the propagation of electromagnetic waves, should be described in terms of mechanistic models. This preference was rooted in a conviction that mechanics was the most basic science in terms of which all phenomena could be analyzed.

Scientists' preferences among models would have little effect on theory choice if any theory could be reformulated to invoke a model of a given kind. If this were so, then every scientist, while cherishing models of a particular kind, could be equally content with all theories. In fact, theories that invoke a particular model cannot easily be transformed so as to invoke some other model: a model is a relatively unreplaceable component of a theory. This means that a scientist's preference for one kind of model over another can indeed function as a determinate criterion for theory choice.

The distinction that I drew earlier between properties and their classes helps clarify this point about models. I interpret "Theory T invokes some model or other" on a par with "Theory T shows some form of symmetry or other": both propositions refer to classes of properties rather than to properties themselves. In contrast, "Theory T invokes a model of such-and-such a kind" (such as a mechanistic model) and "Theory T shows such-and-such a form of symmetry" are descriptions of the theory in terms of its properties, which may be the grounds of scientists' evaluations of T.

Scientists' preferences regarding models exhibit an interesting phenomenon: the kinds of models that scientists require theories to invoke change slowly over time. For instance, the requirement that theories should invoke mechanistic models was held widely by physicists in the nineteenth century but is not generally regarded as appropriate today. Such shifts can be interpreted as the effect of fashions. According to Ernest Nagel, for instance, scientists tend to regard a theory highly if it appeals to a familiar model and penalize it if it invokes an unfamiliar one: "It is a matter of historical record that there are fashions in the preferences scientists exhibit for various kinds of models [. . .]. Theories based on unfamiliar models frequently encounter strong resistance until the novel ideas have lost their strangeness, so that a new generation will often accept as a matter of course a type of model which to a preceding generation was unsatisfactory because it was unfamiliar."[18] As we are about to see, case studies in history of science corroborate Nagel's observation about fashions in model preference. These studies also reveal a pattern underlying such fashions.

Theories in physiology since the sixteenth century have been dominated by models drawn from physics and engineering. At any one time, physiologists have widely agreed about which physical theories constitute the most appropriate source of models, but have disagreed with their predecessors over which theories these were. Physiologists have tended to describe human beings in terms of the most successful physical theories of their own epoch. Mary B. Hesse attributes to Norbert Wiener the observation that "there have been three stages in the scientific description of human beings according to what was the most typical machine in use during the period—first, in the seventeenth and eighteenth centuries, clockwork mechanisms described by analogies from dynamics; then in the nineteenth century, heat engines described by analogies from thermodynamics; and now communication devices described by analogies from electronics."[19] The theories of the seventeenth and eighteenth centuries that had accumulated the greatest empirical glory were, thanks to Descartes and Newton, theories in mechanics. Accordingly, physiology in this period modeled organisms as arrangements of rods, wheels, and cords; and it evaluated new theories according to whether they admitted or failed to admit such models. By the nineteenth century, the empirical success of mechanics had to some extent been eclipsed by that of theories in thermodynamics: physiology's preferred source of models changed in response. In the twentieth century, new patterns of empirical success have prompted physiology to look to electronic engineering for its models.

A similar succession of models is to be found in one of the younger branches of physiology, neurophysiology. Here, too, models have been sought from the physical sciences, but the neurophysiologists' preferred source of models has changed repeatedly. Neurophysiologists in the 1940s were accustomed to likening the nervous system to a telephone switchboard, by inspiration from information theory. This model prompted the idea that the nervous system has a level of organization above that of electrical nerve impulses, at which transfers of information take place; it also suggested that the nervous system is a passive network,

18. Nagel (1961), p. 115.
19. Hesse (1954), p. 140; Hesse refers the reader to Wiener (1948), pp. 39–40. For further discussion of mechanical models of human beings in the seventeenth and eighteenth centuries, see McReynolds (1990), pp. 152–158, and Channell (1991), pp. 30–45.

The Aesthetic Properties of Scientific Theories

in which no response arises if no stimulus is received. In the 1950s, the nervous system was interpreted as a feedback mechanism like a thermostat, by inspiration from cybernetics. This encouraged a view of the nervous system as an essentially active system which constantly counteracts deviations from chemical equilibrium. In the 1970s, neurophysiologists took to likening the nervous system to a central processing unit, by inspiration from computer science. This analogy suggested that the brain possesses structures of the kind necessary for handling "programs": some evidence was found that each cortical cell in the brain is not an individual detector of stimuli, with a unique selectivity, but rather has multiple selectivities and is a member of a network of cells acting jointly as a detector.[20]

It seems that, in neurophysiology as well as physiology, scientists at each time have been choosing as the source of their models the theory or field of the physical sciences that had in the immediately preceding period shown the greatest or at least the most eye-catching empirical success. Chapter 5 will throw more light onto how models succeed one another in a community's preferences. My final claim here is that the property of a theory of invoking a model of a particular kind should be considered an aesthetic property. Analogical reasoning aims to reveal the presence of unity in diversity, a typically aesthetic concern; and metaphors and analogies are sources of aesthetic pleasure in works of literature and art.[21] It can hardly be in doubt that the discovery that a theory lends itself to interpretation through an analogy of a favored kind yields aesthetic satisfaction to a beholder.

4. VISUALIZATION AND ABSTRACTNESS

To carry out their function of accounting for bodies of empirical data, some theories postulate visualizable structures or mechanisms that are said to underlie phenomena. By visualizable structure or mechanism, I mean one of which there is a mental image, drawn typically from everyday experience, that guides our understanding of the nature or dynamics of the phenomena.[22]

Here are some examples of visualizable structures and mechanisms

---

20. On these phases in the history of neurophysiology, see Pribram (1990), pp. 81–88.
21. A study of the features common to the use of analogy in science and in literature is Beer (1983), pp. 79–103.
22. For an introduction to the role of visualization in scientific thinking, see Arnheim (1969), pp. 274–293. Scientists' use of visual thinking is documented in Shepard (1978), pp. 125–127, and Root-Bernstein (1985), pp. 52–58.

postulated by scientific theories. In the phlogiston theory in chemistry, the combustion of a body was visualized as involving the release of a fluid, phlogiston. Since Hermann von Helmholtz, non-Euclidean space has been pictured as a two-dimensional surface curved in the third dimension.[23] The "spin" of an elementary particle such as an electron is sometimes visualized as a rotation about an axis.

It commonly occurs that different theories offer different visualizations of a phenomenon. For example, the interaction of two electrons as they approach one another is visualized by classical electromagnetic theory as the gradual intensifying of a repulsive electrostatic force, and by quantum electrodynamics as the exchange of a virtual photon. More remarkably, some theories visualize a phenomenon in more than one way. For example, the standard present-day theory of nuclear magnetic resonance—which occurs when a material is exposed to a magnetic field oscillating at a particular frequency—suggests two visual images. According to the first visualization, a material's atomic nuclei absorb energy from the magnetic field and undergo transitions from one quantum state to another; according to the second visualization, the oscillating magnetic field repeatedly reorients the magnetic moments of the nuclei.[24]

Other theories give accounts of empirical data that are not pictorial but abstract. An abstract theory does not evoke a mental image: rather, it describes phenomena by means solely of a mathematical or other formal apparatus. For example, mechanics in the century after Newton was developed largely in an abstract style, not depending on particular visualizations. In the preface to his compendium on the subject, titled *Mécanique analytique*, Joseph Louis Lagrange declares, "No figures will be found in this work. The methods that I here set forth require neither constructions nor geometrical or mechanical arguments, but only algebraic operations, subject to a regular and uniform procedure."[25]

Visualization and abstractness are among the properties on the strength of which scientists evaluate and choose between theories. Many scientists declare a preference for visualizing theories: Einstein and Minkowski were among these.[26] So was Richard P. Feynman; indeed, he is perhaps best known for developing what are now called Feynman diagrams, pictorial representations of certain quantum-mechanical interactions between elementary particles.[27]

23. Helmholtz (1870), pp. 5–11.
24. The two visualizations of nuclear magnetic resonance are described in Rigden (1986).
25. Lagrange (1788), pp. xi–xii.
26. On Einstein's predilection for visualization, see Holton (1973), pp. 385–388; on Minkowski's, see Galison (1979).
27. On Feynman's use of visualization, see Schweber (1994), pp. 462–467.

The Aesthetic Properties of Scientific Theories

But it is not the case that every scientist cherishes visualization: some prefer abstract reasoning. In the case of some scientists, the latter preference is motivated by positivism. Scientists who are positivists tend to believe that one ought to advance only claims for which there is sufficient empirical evidence, and that such claims take the form of mathematical—and thus abstract—relations between observable magnitudes. To try to complement an abstract theory with visualizable mechanisms is, according to them, to go beyond what is warranted.

Visualization and abstractness should not be thought of as inessential or eliminable qualities of theories, any more than invoking a model is. Here are two arguments that suggest that, on the contrary, they are deep-seated properties, characteristic of particular theories. First, an abstract theory generally cannot be reformulated so as to refer to visualizable processes and remain recognizably the same theory: there is often simply no mental image, whether drawn from everyday experience or elsewhere, that can depict the relations between physical variables that the theory posits. For instance, many of the theories found in present-day submicroscopic physics not only originated in a nonvisual form but also have shown themselves refractory to subsequent attempts to find convincing visualizations for them. Second, a theory that refers to a visualizable mechanism cannot generally be reduced to a purely abstract theory without the loss of some explanatory or heuristic power: it frequently turns out that the visualization that such a theory puts forth plays a role in generating the theory's explanations of phenomena or in showing how the theory should be applied or further developed.

The property of theories of being visualizing should not be confused with the property of invoking a model. True, a visualizing theory posits a relation between the phenomena that it describes and a concrete mechanism in some other domain of experience: and this relation is similar to one of analogy. Nevertheless, there are two facts that require us to distinguish the property of visualization from the property of invoking a model.

First, as we saw in the previous section, model-based reasoning involves a transfer of a stock of conceptual and analytical resources between distinct scientific domains. But requiring that a mechanism postulated by a theory be visualizable stops short of demanding that it be possible to perform such a transfer of resources to the theory. When we say that the wave theory of light invokes the model of waves in water, we allude to the fact that a system of concepts and relations is transferred from the theory of water waves to the theory of light. This includes the concepts of wavelength, frequency, amplitude, diffraction, and interference; the relation of proportionality linking velocity, wavelength, and frequency; and the wave equation. By contrast, when we claim that space in a finite, unbounded, expanding universe can be visualized as the surface of an inflating balloon, we do not thereby imply that cosmology acquires conceptual resources from theories of balloons; when we visualize a strand of DNA as a spiral staircase, there is no benefit that we expect theories in biochemistry and genetics to draw from civil engineering.[28]

The second fact obliging us to distinguish visualization and appeal to models is that there is nothing about the notion of a good model that requires it to be visualizable. The demand that a mechanism be visualizable is a demand for a mechanism that is pictorially representable: while this mechanism generally exists nowhere but in scientists' imagination, it can in principle be depicted in diagrams, or even manufactured. By contrast, a satisfactory model can be provided by a wholly abstract formalism. For instance, present-day particle physics uses some models drawn from a branch of mathematics known as group theory, which describes the properties of transformations. According to a highly regarded theory put forward by Howard Georgi and Shelley Glashow in 1974, the structure of the classification of quarks and leptons is isomorphic to the structure of a particular group, called $SU(5)$.[29] Although this theory invokes a model, it cannot be considered thereby to be a visualizing theory, since the model that it invokes is abstract. This shows that to provide a model by which a particular theory can be interpreted does not ensure that the mechanisms that the theory postulates are visualizable.

Perhaps the best way of drawing the distinction between the property of invoking a particular model and the property of visualization is to say that whereas a theory that invokes a model appeals to a relation of analogy, a theory that offers a visualization of a phenomenon constructs a relation of metaphor. After all, to draw an analogy is to point out a homology between two structures, which is just what a model relies on; and to use a metaphor is to see something as something else, which is what a visualizing theory prompts us to do.

The distinction between the property of using a model and the property of offering a visualization has frequently been overlooked by philosophers of science. For instance, Hesse identifies two schools of thought in physics, which she retraces to Norman R. Campbell and Pierre Duhem. In Hesse's account, Campbell asserts and Duhem denies that

28. The visualization of the expanding universe as the dilating surface of a balloon is suggested for example by Hoyle (1950), pp. 102–103.

29. The use in particle physics of models constituted by symmetry groups is described in Zee (1986), pp. 228–254.

models play an essential role in scientific theorizing. However, some of the statements that she cites as illustrations of this dispute are appeals to visualizations rather than to models. For instance, Hesse portrays followers of Campbell and Duhem as arguing whether August Kekulé's dream of a snake gripping its tail in its mouth played an essential role in his conjecturing that the benzene molecule has the structure of a ring of six carbon atoms.[30] But the snake dreamt by Kekulé does not, in any tenable sense, function as a model: it does not effect a transfer of concepts or relations from herpetology to structural chemistry. The snake is rather a visualization of the structure that Kekulé's theory attributes to the benzene molecule. Hesse is correct in detecting many appeals to models and analogies in science, but to portray a visualization as a model is to trivialize the latter notion.[31] Much of the dispute between Campbell and Duhem is better interpreted as one between a visualizing style of theorizing typical of British physics and an abstract style favored by French physicists. More particularly, Duhem's well-known criticism of what he called the English school of physics is directed as much against their insistence on providing visualizations of phenomena as against the use of models.[32]

I claim that the properties of visualization and abstractness are aesthetic properties of theories. A theory's suggestion of a certain visualization for a structure or mechanism is liable to raise in some beholders a sense of aptness, which satisfies the second criterion that I proposed in Chapter 2 for recognizing aesthetic properties of theories. Other scientists see aesthetic value in abstract theories, deriving pleasure from regarding a pure, formal conceptual structure whose power does not depend on a particular pictorial interpretation. Duhem took aesthetic pleasure in such theories: "It is impossible to follow the march of one of the great theories of physics, to see it unroll majestically its regular deductions starting from initial hypotheses, to see its consequences represent a multitude of experimental laws down to the smallest detail, without being charmed by the beauty of such a construction, without feeling keenly that such a creation of the human mind is truly a work of art."[33] In contrast, visualizing theories displeased him because they appeal to the imagination. For the English physicist, Duhem complains, "Theory is [. . .] neither an explanation nor a rational classification of

---

30. Hesse (1966), p. 7.

31. Hesse's neglect of the distinction between models and visualizations is criticized by Mellor (1968), pp. 282–285. Nagel (1961), pp. 107–117, similarly confuses visualization with reference to a model.

32. Duhem (1906), pp. 69–104.

33. Ibid., p. 24.

physical laws, but a model of these laws, a model not built for the satisfying of reason but for the pleasure of the imagination. [. . .] Thus, in English theories we find those disparities, those incoherencies, those contradictions which we are driven to judge severely because we seek a rational system where the author has sought to give us only a work of imagination."[34] Duhem's aesthetic distaste here is obvious.

Yet another reason why the property of visualization is of interest for our discussion is that its instantiation is largely nonmathematical; that is, this property does not manifest itself in the mathematical structure of theories. In this respect, the property of visualization differs from other aesthetic properties of theories, such as their forms of symmetry. It is often remarked that a theory's symmetry is best revealed when the theory is couched in a mathematical formalism. Some have concluded from this that the beauty of empirical theories is nothing other than mathematical beauty. But visualizability, as we have seen, depends frequently on a theory's placing less rather than more reliance on a mathematical formalism: indeed, putting great emphasis on formalism can lead to a loss of visualization.

The existence both of scientists who prefer visualizing theories and of scientists who prefer abstract theories has given rise to an interesting phenomenon of theory succession. There have been several historical periods in which the two preferences have competed in the community's canon for theory choice. This competition has had one of two effects. In some cases, the community involved has divided into two factions, one of visualizing and one of abstracting scientists, each holding to its own theories. In other cases, the community has swung from a visualizing theory to an abstract theory or vice versa.

The latter outcome is exemplified by the development of quantum theory from 1913 to 1927, a period in which leading theories of subatomic particles lost visualization; we shall look at that episode in detail in Chapter 11. The former outcome, in which factions of visualizing and abstracting scientists coexist in a community, is exemplified in nineteenth-century electrodynamics. Mainly through the work of Siméon-Denis Poisson, physicists had developed techniques to describe electrostatic and magnetic fields mathematically. The mathematical treatment attributed no property to these fields other than the potential, that is, the power to exert forces on electrically charged bodies placed within them. This suggested that charged bodies and magnets act upon one another at a distance, rather than through an intervening medium. To those of an

34. Ibid., p. 81. I interpret Duhem in this passage to be speaking of visualizations, despite the use of the word "model" in the translation.

empiricist bent, such as Michael Faraday and other British physicists, this idea was unsatisfactory: they wished to picture events in electric and magnetic fields more concretely. Faraday did not seek to amend Poisson's mathematical treatment but complemented it with a technique to picture electromagnetic fields as regions of space permeated by "lines of force" emanating from charged bodies and magnets. This visualization of the field partly inspired Maxwell's later extension of electromagnetic theory. In the following passage, Maxwell describes the difference between the theory incorporating Faraday's visualization of electromagnetic fields and the purely abstract treatment given by the "mathematicians," as he calls them: "Faraday, in his mind's eye, saw lines of force traversing all space where the mathematicians saw centres of force attracting at a distance: Faraday saw a medium where they saw nothing but distance: Faraday sought the seat of phenomena in real actions going on in the medium, they were satisfied that they had found it in a power of action at a distance impressed on the electric fluids."[35]

Faraday's contribution ensured that in the second half of the nineteenth century there were two distinct versions of electromagnetic theory: one purely mathematical and abstract, which made no pronouncements about media of propagation and suggested that electromagnetic interactions occurred by action at a distance, and one that, though sharing the mathematical formalism of the first, went on to supply a visualization of fields in terms of lines of force. Between these two theories, physicists were free to choose on the basis of their preferences for abstractness or visualization.[36]

5. METAPHYSICAL ALLEGIANCE

Each of the great metaphysical world views that is recorded in intellectual history is a complex entity. One of its components is a set of claims about the ultimate constituents of the world; a second is a set of norms of reasoning; a third is a set of prescriptions stipulating which sorts of empirical claims about the world should be entertained and which rejected. For instance, atomism describes the world as composed of mate-

35. Maxwell (1873), 1:ix. For historical material on Faraday's visualization of the electromagnetic field, see Hesse (1961), pp. 198–203; on Maxwell, see Kargon (1969), pp. 431–436. For further discussion of Faraday's and Maxwell's use of imagery, see Nersessian (1988).

36. A choice between visualizing and abstract theories presented itself also in nineteenth-century German electrodynamics, as Caneva (1978) shows: see pp. 68–70 on what he calls "concretizing science," and pp. 95–104 on "abstracting science."

rial corpuscles; it prescribes that the properties of macroscopic bodies should be explained by appeal to the properties of their constituent atoms; and, since it interprets the propagation of light as a stream of corpuscles, it entails a rejection of the empirical claim that light propagates instantaneously.

It is this third component of metaphysical world views, the set of criteria for the acceptability of empirical claims about the world, that is of interest here. This component amounts to a set of metaphysical criteria for the evaluation of scientific theories. Different theories exhibit, by virtue of their claims, allegiances to different metaphysical world views. A scientist who holds to a particular metaphysical world view may thus evaluate theories partly according to the metaphysical allegiances that they exhibit.[37]

I propose to regard the allegiances that scientific theories have to metaphysical world views as aesthetic properties of them. This is not the customary approach: philosophers of science more usually regard aesthetic preferences in theory evaluation as a subset of metaphysical preferences, perhaps in the belief that scientists' aesthetic tastes are shaped by their metaphysical outlook.[38] My justification for inverting the usual classification is that the property of having a particular metaphysical allegiance strongly resembles the other properties of theories described in this chapter, in two regards.

First, a beholder who perceives an accord between the claims of a given scientific theory and his or her metaphysical commitments is likely to experience a sense of aptness. Conversely, a theory whose metaphysical allegiance conflicts with the convictions of the beholder will elicit distaste. The usual evidence for this claim is Einstein's reaction to quantum theory, which we shall examine in Chapter 11.

Second, the procedure by which scientific communities form and update the metaphysical criteria on which they judge theories is identical to the procedure by which they choose the form of symmetry, family of models, or degree of visualizability that they will favor in theory evaluation. As their metaphysical criteria, scientific communities choose those that would have been satisfied by their empirically most successful theories of the recent past. Embracing such criteria has two effects. It strengthens the conviction that those theories were indeed worthy of high esteem; and it encourages scientists to seek further theories that

37. For a general discussion of the role of metaphysical criteria in theory assessment, see Agassi (1964).

38. Classifications of aesthetic criteria as a subset of metaphysical criteria are given by Margenau (1950), p. 81, and Buchdahl (1970), p. 206.

The Aesthetic Properties of Scientific Theories

show the same metaphysical allegiance. This means that a set of metaphysical presuppositions will tend to entrench itself increasingly.

Here is an illustration of my claim that, as their metaphysical criteria, scientists select criteria that are fulfilled by their empirically most successful theories of the recent past. It pertains to the rise and fall in the seventeenth century of the criterion that theories ought not to attribute to inanimate objects such properties as active powers, occult qualities, and a capacity for action at a distance.[39]

During the Renaissance, in the effort to explain the interactions of bodies, some protoscientific disciplines such as astrology, alchemy, and magic attributed active powers to inanimate matter. An active power is the capacity to originate influences that affect other entities, rather than just to transmit influences that have been originated elsewhere. According to the theories put forward in these disciplines, possession of active powers was an imperceptible or occult quality of inanimate entities. Moreover, these theories maintained that the influences arising from entities' active powers were propagated at a distance, i.e., with neither contact between the originating and the receiving entity nor activity in an intervening medium.

For example, astrology attributed to heavenly bodies a capacity to influence human affairs at a distance. Alchemy hypothesized the existence of microcosm–macrocosm correlations through which, for instance, particular substances had a medicinal effect in virtue of standing in a certain relation to the universe. Magical theories attributed to preparations of herbs and minerals the power to affect people and objects from afar. According to an enduring belief, for example, certain wounds could be healed by applying a salve to the weapon that had inflicted them. Magic invariably assumed that the qualities in virtue of which substances possessed these active powers were occult.[40]

Corpuscularism, which arose during the seventeenth century from the work of René Descartes, Pierre Gassendi, Robert Boyle, and others, was suspicious of the notions of active power, occult quality, and action at a distance. It aimed to eliminate these notions from natural philosophy. It wished to explain all phenomena by appeal to corpuscles that do not originate influences of their own, that possess no nonperceptible qualities, and that interact with other particles only by contact and impact, through which they acquire their motions. This program reached its fullest realization in Cartesian physics. For example, Descartes interpreted

39. The history of natural philosophers' attitudes towards action at a distance is discussed by Hesse (1961), pp. 98–188, and Buchdahl (1973), on whose accounts I draw.
40. On the occult disciplines of the Renaissance, see Webster (1982).

electrostatic and magnetic forces as the effects of differential pressures exerted by streams of particles. The fall of objects near the earth's surface was explained by supposing that the earth's rotation produces a centrifugal force that has a differential effect on different substances: earthy particles fall to compensate for the rise of other particles. Planets are swept round in their orbits by huge vortices of particles with which the universe is filled, while comets are carried from one vortex to another. In these theories, the notion of action at a distance was both unnecessary and unintelligible.

Appreciation of Kepler's laws of planetary motion and the sophistication of his mathematical techniques enabled Newton to formulate a quantitative, testable theory of universal gravitation. This theory asserted that each particle of matter in the universe emanates a force attracting every other particle: the force acts at a distance, without activity in any intervening medium. Initially, Newton searched for a corpuscular mechanism that might account for this attraction; but this search was unsuccessful, and ultimately he offered no corpuscularist explanation of gravity. He also appealed to attractive and repulsive forces acting at a distance in his theories of the reflection of light, the cohesion of solids, and chemical reactions. He wrote that all such forces are the manifestation of active powers that are intrinsic to matter.

In other words, Newton readmitted into physical theory concepts that would have been familiar to Renaissance practitioners of astrology, alchemy, and magic. The fact that some of Newton's concepts recall occult theories may not be fortuitous: it appears that Newton's thinking in natural philosophy was influenced by a strong interest that he maintained in alchemy.[41] Cartesian natural philosophers such as Christiaan Huygens and Gottfried Wilhelm Leibniz criticized Newton for appealing to these concepts. They saw his theory as returning to the occultism and mysticism which corpuscularism had intended to eliminate. There followed a lengthy dispute between the Newtonians and the Cartesians over whether a theory could and should offer corpuscularist explanations of gravitational phenomena.[42] Nonetheless, in the eighteenth century the theory of action at a distance took hold among physicists even in France, where Descartes's prestige was greatest. Soon, far from being considered unintelligible, the active power of gravity was viewed as a wholly unexceptionable property of matter.

Between the late sixteenth and the late eighteenth centuries, therefore,

41. Newton's interest in alchemy is documented by Dobbs (1992).
42. On the dispute between the Newtonians and the Cartesians over gravity, see I. B. Cohen (1980), pp. 79–83, and Hutchison (1982), pp. 250–253.

The Aesthetic Properties of Scientific Theories

there occurred two great changes in the criteria for theory evaluation in natural philosophy. During the first change, which accompanied the rise of Cartesian physics, it became a requirement that theories should analyze phenomena in terms of impacts among inactive corpuscles, and avoid reference to active powers, occult qualities, and action at a distance. In the second change, which accompanied the rise of Newtonian physics, the ban on the use of such notions in natural philosophy was relaxed, and some of them came to be highly valued.

The question of interest here is, what prompted these successive changes? Among the relevant factors are the relative degrees of empirical success of the available theories that satisfied the alternative sets of criteria. The Renaissance astrological, alchemical, and magical theories were perceived ultimately to have been unsuccessful at their explanatory and predictive tasks. Partly in the light of the unsatisfactory empirical record of these theories, natural philosophers came to doubt the value of their metaphysical assumptions. In contrast, corpuscularist theories such as those of Descartes scored encouraging empirical successes in several areas, including investigations of light, heat, fluids, and planetary motions. Their success ensured that the metaphysical allegiances of such theories came to be better regarded. This esteem was reflected in the requirement that came to be placed on theories during the second half of the seventeenth century, that they should show corpuscularist allegiances. Newton himself initially felt the influence of this requirement, searching for a corpuscularist explanation of gravity. His failure to find one did not, however, prevent him from putting forward a non-corpuscularist theory that demonstrated empirical success. Cartesians, among whom corpuscularism was more deeply entrenched, rejected Newton's theory. With the passage of time, Newton's theory accumulated empirical success far greater than that of Cartesian physics. This success ensured that the metaphysical allegiances of Newtonian theories came to be accepted by the community, even though they had been shared by some protoscientific accounts of nature. The community came to embrace criteria for theory evaluation that can fittingly be called Newtonian, since they were shaped by properties of Newton's theories. These criteria influenced theorizing and theory choice in physics until the end of the nineteenth century.

A similar story can convincingly be told about, for example, the decline of the requirement that physical theories should be deterministic, which began in the 1920s: here too the succession of metaphysical criteria was driven partly by considerations of empirical success. In this respect, the behavior of scientists' metaphysical criteria resembles that of the other criteria that we have been examining in this chapter. There is therefore some hope that the formulation and updating of all such criteria may be described by the same model.

6. Beauty in the Biological and Social Sciences

Aesthetic factors seem to feature more prominently in the methodological remarks of physical scientists than in those of biological and social scientists. This disparity has fostered in some writers the impression that aesthetic factors operate mainly in the physical sciences and very little in other sciences. Although this book refers more extensively to physical science as a source of historical examples, I see evidence that aesthetic factors have great influence in all branches of science. The model that I shall construct of the development of scientists' aesthetic canons applies to the biological and social sciences as much as to the physical sciences. For instance, I claim that the aesthetic induction (Chapter 5) operates throughout the sciences and that my model of scientific revolutions (Chapter 8) fits all branches of science.

Of course, the properties of theories that are attributed aesthetic value in the physical sciences may differ from those that are attributed it in the biological and social sciences. Here are two illustrations.

First, since it is much more common for typical theories in the physical sciences than for those in the biological and social sciences to have an explicit mathematical structure, the conception of beauty that physicists have tends to refer much more heavily than that of biologists and social scientists to formal and mathematical properties of theories, such as symmetries. Indeed, physicists frequently describe their theories as having "mathematical beauty"—a phrase not commonly used about theories in the biological and social sciences. Now, the grounds on which beauty is attributed to constructs in pure mathematics, such as theorems and proofs, may differ from the grounds on which it is attributed to theories in empirical science, since the latter can refer to measures of empirical success while the former cannot. Because of this, it may be that a philosophical treatment of beauty in empirical science, such as the one given in this book, does not apply to beauty in pure mathematics.[43] Nonetheless, mathematical structure clearly contributes to determine the attributions of aesthetic value to physical theories.

Second, practitioners of the physical sciences hold the simplicity of theories in higher regard than do biological and social scientists. Differ-

43. Among well-known treatments of the beauty of mathematical constructs are Le Lionnais (1948) and Huntley (1970).

The Aesthetic Properties of Scientific Theories

ent sciences have, to some extent, different aims. Typical theories in physical science aim to formulate laws of nature, universal generalizations that are frequently only approximately true but are invariably concise. Regard for laws of nature can easily lead a physicist to believe that the most striking property of theories is their simplicity relative to the range of phenomena that they describe. A typical theory in the biological sciences aims to formulate not laws of nature but rather a detailed and differentiated account of comparatively few phenomena. Biological scientists therefore are not led to value theories that are notably simple.

But this does not mean that there are no properties of theories to which aesthetic value is attached in the biological and social sciences. For instance, the consideration that a particular theory lends itself to treatment by an analogy, or offers a visualization of phenomena, or has a particular metaphysical allegiance plays just as important a role in the biological and social sciences as in physical science.[44]

The high visibility that physics enjoys in philosophical discussions of science has led some writers to conclude that the sole aesthetic properties of theories that scientists recognize are formal properties, such as their symmetry and simplicity properties, rather than properties relating to the content of theories. These writers have put forward formalist accounts of the aesthetic properties of theories.[45] I believe that a formalist approach will not yield an adequate understanding of scientists' aesthetic responses to theories, since other aesthetic properties that scientists recognize pertain—as we have seen in this chapter—to the content of theories rather than their formal structure.

44. Previous literature on the role of aesthetic factors in the formulation and assessment of theories in the biological sciences includes Ghiselin (1976), Pickvance (1986), pp. 150–153, and Root-Bernstein (1987); on social science, see Nisbet (1976), pp. 9–26.

45. A formalist treatment of the aesthetic properties of scientific theories is advanced by Engler (1990, 1994).

Two Erroneous Views of
Scientists' Aesthetic Judgments

1. The Theory of Aesthetic Disinterestedness

If it is true that scientists appraise theories both on empirical criteria, such as those proposed by the logico-empirical model of theory evaluation, and on aesthetic criteria, such as those reviewed in the previous chapter, what relation holds between appraisals of the two sorts? In this chapter, we examine two views of this relation and gauge their adequacy in the light of scientists' testimony and other evidence from the history of science. These views will prove inadequate, but our examination of them will offer clues toward a more adequate view to be developed in the next chapter.

The two views that we shall examine are autonomism and reductionism. Autonomism regards scientists' aesthetic and empirical evaluations as wholly distinct from and irreducible to one another, whereas reductionism views them as nothing but aspects of one another. These represent the extremes of a spectrum of possible views, each of which posits a certain degree of interreducibility between aesthetic and empirical judgments. We shall first examine the autonomist view.

Here is a theory about the mode of attention that is activated in aesthetic perception.[1] There are many modes of attention that one may adopt in perception. These modes are characterized by the aims or interests by which the perception is animated or in view of which it is conducted. One may for instance gaze upon a gem to make a valuation of it,

---

1. I have previously investigated this model of scientists' aesthetic judgment in McAllister (1991a).

or one may survey a chessboard with the interests of Black at heart. According to the theory set out here, the mode of attention that is characteristic of aesthetic perception is disinterested. "Disinterestedness" denotes an attitude of detachment or purposelessness toward an object of perception. The disinterested stance toward an object dwells not upon the object's aptitude to further some aim but only upon its intrinsic structure and significance.

The notion of perceptual disinterestedness acquired prominence in eighteenth-century moral philosophy. According to Lord Shaftesbury, moral rectitude is attained not, as Thomas Hobbes maintained, through the cultivation of one's interests but by dissociating one's conduct from all interests and searching for propriety in moral acts. The mind discerns which acts are proper by operating in an attitude of disregard for interests.[2] But if the mind is capable of operating thus when seeking out morally righteous acts, it may do so in aesthetic perception as well. Indeed, Shaftesbury identifies aesthetic perception as the mode of perception that has no regard for ulterior interests: "Imagine [. . .] if being taken with the beauty of the ocean, which you see yonder at a distance, it should come into your head to seek how to command it, and, like some mighty admiral, ride master of the sea, would not the fancy be a little absurd?"[3] The fulfillment that would derive from "possessing" the ocean is "very different from that which should naturally follow from the contemplation of the ocean's beauty."[4] This is because the aesthetic contemplation of an object attributes no utilitarian dimension to it.

Hutcheson agrees with Shaftesbury that aesthetic judgment pays no regard to utilitarian concerns. He believes this to be demonstrated by the observation that we are often disposed to pursue the beautiful to our cost: "Do not we often see convenience and use neglected to obtain beauty, without any other prospect of advantage in the beautiful form than the suggesting the pleasant ideas of beauty?"[5]

The suggestion that this mode of attention may be applied to scientific theories occurs when Shaftesbury turns from the perception of natural beauty to the aesthetic pleasure caused by theorems in mathematics:

There is no one who, by the least progress in science or learning, has come to know barely the principles of mathematics, but has found, that

2. See Shaftesbury (1711), 1:251; compare Hutcheson (1725), p. 25.
3. Shaftesbury (1711), 2:126. The centrality of the notion of aesthetic disinterestedness to Shaftesbury's thought and to the rise of modern aesthetics is emphasized by Stolnitz (1961a, 1961b); the continuing controversy provoked by Stolnitz's interpretation is reviewed by Arregui and Arnau (1994).
4. Shaftesbury (1711), 2:127.
5. Hutcheson (1725), p. 37. On aesthetic disinterestedness in Hutcheson, see Kivy (1976), pp. 73–75.

in the exercise of his mind on the discoveries he there makes, though merely of speculative truths, he receives a pleasure and delight superior to that of sense. When we have thoroughly searched into the nature of this contemplative delight, we shall find it of a kind which relates not in the least to any private interest of the creature, nor has for its object any self-good or advantage of the private system.[6]

Hutcheson similarly distinguishes the aesthetic contemplation of scientific knowledge from an awareness of its utility: "It is easy to see how men are charmed with the beauty of such knowledge, besides its usefulness [. . .]. And this pleasure we enjoy even when we have no prospect of obtaining any other advantage from such manner of deduction than the immediate pleasure of contemplating the beauty."[7] Shaftesbury and Hutcheson thus suggest that mathematical theorems and scientific theories may be evaluated on two scales: one measuring the degree of utility that the constructs afford, and the other relating to the aesthetic pleasure that is obtained by contemplating them disinterestedly.

The theory of aesthetic disinterestedness has been revived by some twentieth-century writers on aesthetics. Edward Bullough asks us to imagine our sensations upon being caught in a sea fog.[8] In ordinary circumstances a fog at sea is the cause of anxiety: mariners are likely to fear for the safety of their vessel. "Nevertheless," Bullough writes, "a fog at sea can be a source of intense relish and enjoyment. Abstract from the experience of the sea fog, for the moment [. . .]; direct the attention to the features 'objectively' constituting the phenomenon."[9] Bullough suggests that when this effort is made, the attention of the observer is absorbed by the fog's milky veil, the carrying power of the air, the creamy smoothness of the water, the feeling of remoteness from the world. When perception is directed to such features for their own sake rather than out of concern for the dangers that they pose to navigation, "the experience may acquire [. . .] a flavour of such concentrated poignancy and delight as to contrast sharply with the blind and distempered anxiety of its other aspects."[10] This difference of outlook is due to the observer's dissociating from utilitarian interests and exercising a mode of disinterested attention. Bullough speaks of this abstraction as the interposition of "psychical distance" between the observer and the object of contemplation. "Distance [. . .] is obtained by separating the object and its appeal from

6. Shaftesbury (1711), 1:296.
7. Hutcheson (1725), p. 51.
8. Bullough (1912), pp. 87–118.
9. Ibid., p. 88.
10. Ibid., p. 89.

Two Erroneous Views of Scientists' Aesthetic Judgments

one's own self, by putting it out of gear with practical needs and ends. Thereby the 'contemplation' of the object becomes alone possible."[11]

Two modes of perception may thus be identified, corresponding to two sets of values that objects may possess. The values of one set are utilitarian, whereas the others—perceived via the interposition of psychical distance—are aesthetic: "Distance [. . .] supplies one of the special criteria of aesthetic values as distinct from practical (utilitarian), scientific, or social (ethical) values. All these are concrete values, either *directly* personal as utilitarian, or *indirectly* remotely personal, as moral values."[12] We may regard any object under both modes of perception, attributing it both a degree of practical value in an act of utilitarian perception and a degree of aesthetic value in the course of disinterested perception.[13]

The theory of aesthetic disinterestedness has been defended more recently by Jerome Stolnitz. He characterizes the "aesthetic attitude" as "disinterested and sympathetic attention to and contemplation of any object of awareness whatever, for its own sake alone," with "no concern for any ulterior purpose."[14]

Bullough does not explicitly consider scientific theories as objects of aesthetic perception; Stolnitz suggests that disinterested aesthetic judgment is exercised in mathematics, but he does not mention the empirical sciences as a possible scene of aesthetic judgment.[15] Nonetheless, the views of these writers can easily be extended to the perception of scientific theories. Bullough distinguishes between appreciation of an object's aesthetic values and of its "scientific values." The "scientific values" of scientific theories might be interpreted as their utilitarian power, revealed in empirical tests and practical applications. On this interpretation, Bullough is implicitly suggesting that scientific theories may be judged on two sets of criteria. One is utilitarian, composed of logical and empirical criteria suited to ascertain theories' empirical worth; the other is an aesthetic canon paying no regard to utilitarian virtues.

2. The Accord of Aesthetic and Empirical Judgments

According to the theory of aesthetic disinterestedness, as we have seen, the evaluations of a theory on empirical and on aesthetic criteria are inde-

---

11. Ibid., p. 91.
12. Ibid., pp. 117–118.
13. Criticism of Bullough's theory of disinterested attention is advanced by Dickie (1974), pp. 91–112, and K. Price (1977), pp. 411–423.
14. Stolnitz (1960), pp. 34–35.
15. Ibid., pp. 41–42. King (1992), pp. 194–208, suggests that the beauty of mathematical constructs becomes perceptible to us when we take psychical distance from them.

pendent of one another. A theory's measure of empirical success has no correlation with its perceived aesthetic virtue. Let us now evaluate the adequacy of this view as a model of scientists' behavior.

Some scientists indeed distinguish sharply between the empirical properties and the aesthetic properties of theories. Alexander Keller sums up the reception that physicists accorded to the Bohr model of the atom with the words "Very pretty—but will it work?"[16] Joe Rosen argues that one cannot tell from scientists' aesthetic evaluations of theories how highly they rate them on empirical criteria: "If we eavesdrop on private discussions among scientists, we might hear expressions such as, 'This is a beautiful theory (of ours)!' or, 'His theory is rather ugly.' Both theories might be equally good, in that they both explain the same natural phenomena equally well. In fact, the 'ugly' theory might even be better."[17] These statements presuppose that utilitarian and aesthetic evaluations of theories are independent of one another. If the possibility exists that a theory judged ugly is empirically superior to one judged beautiful, it must be that the empirical and aesthetic evaluations are made independently, and a theory's aesthetic worth is assessed without regard for its empirical utility.

However, there is a much larger body of evidence from scientific practice that conflicts with this view. It is necessary to scrutinize more severely the claim that scientists' aesthetic evaluations of theories are independent of their empirical evaluations. It is probably true that there is little correlation between the empirical success of a theory in the first few years following its formulation and the aesthetic value attributed to it at the end of that period. For instance, within twenty years of their formulation in the 1920s, quantum theories of submicroscopic phenomena had accumulated impressive empirical successes but were regarded by many physicists as aesthetically unappealing (see Chapter 11). But in the longer term a correlation between scientists' empirical and aesthetic evaluations tends to emerge. If a theory maintains a good empirical track record for a lengthier period following its formulation, it tends by the end of that period to have attracted scientists' aesthetic approbation as well. For example, by the 1970s the prevailing aesthetic reaction of physicists to quantum theory was no longer antipathy but rather an increasing acknowledgment that the theory had aesthetic virtues of its own, albeit ones different from those of the theories of submicroscopic phenomena that it had superseded.

In other words, it seems that the aesthetic properties of theories that

16. Keller (1983), p. 169.
17. Rosen (1975), pp. 120–121.

Two Erroneous Views of Scientists' Aesthetic Judgments

show persistent empirical success gradually win favor: scientists' aesthetic evaluations tend in the longer term to swing into line with their empirical appraisals. This correlation suggests that aesthetic judgments are in the longer term passed partly in the light of theories' empirical performance. This finding contradicts the model that portrays scientists' aesthetic evaluations as indifferent to empirical worth.

I have suggested that a community's aesthetic judgment of a theory might differ from its empirical evaluation for a few years after its formulation, but is likely to agree with it in the longer term, provided that the theory has continued to demonstrate empirical success. This change is explained by the assumption, developed in the next chapter, that a scientist's aesthetic judgment of an empirically successful theory is reached by a process similar to his or her becoming habituated to relying on the theory. For at least a few years after the formulation of an aesthetically innovative theory, a community's upper ranks will typically continue to be staffed by scientists who were trained to hold to and esteem its predecessors. If the new theory shows unprecedented aesthetic features, the community is likely to find it displeasing. Once a longer time has elapsed, in contrast, the community will have become habituated to its presence. If aesthetic opinions of theories can be attributed to a form of habituation, one may expect the aesthetic judgments that scientists pass at this later time to have swung into accord with their empirical appraisals.

If theories that enjoy persistent empirical success come to be regarded as aesthetically attractive, scientists' aesthetic evaluations of theories cannot be interpreted as disinterested. However, scientists also sometimes pass favorable aesthetic appraisals of theories that are empirically unsuccessful—we shall meet some cases of this in the next section. Is it not plausible to interpret at least these aesthetic judgments as disinterested, as they seem unrelated to judgments of empirical worth? Perhaps, but it is possible to interpret even these aesthetic judgments as correlated with empirical performance, albeit in a more complex way. A theory, while not itself enjoying empirical success, may have aesthetic properties similar to those of other theories that have been empirically successful. If scientists have come to regard these empirically successful theories as aesthetically attractive, they are likely to regard as similarly attractive the empirically less successful theory that shares their aesthetic properties. As an example, imagine a mistaken Newtonian-style theory in fluid mechanics in the late eighteenth century. Although this theory is not successful at accounting for the data in its domain, it shares many of the aesthetic properties of the Newtonian theory in celestial mechanics, which is empirically successful. If Newtonian celestial mechanics is regarded as aesthetically attractive on the basis of its empirical success, then the Newtonian-style fluid mechanics will be regarded as aesthetically attractive despite its lack of empirical success. The aesthetic value that is attributed to it may be unmerited, to the extent that it is due to the empirical success of another theory; but it is nonetheless the outcome of a correlation between aesthetic appreciation and empirical success.

These are the considerations that lead me to reject the theory of aesthetic disinterestedness as an account of scientists' aesthetic judgments of theories. The details of the process by which I shall suggest scientists' aesthetic judgments are formed, and which underlie the remarks in this section, will be laid out in the next chapter.

3. Reductionism about Aesthetic and Empirical Judgments

We have been discussing the view that scientists' aesthetic judgments of theories are autonomous of their empirical judgments. This view marks one extreme of the spectrum of possible views about the relation between aesthetic and empirical judgments. At the opposite extreme is reductionism, which claims that one of these forms of judgment is nothing but a manifestation or aspect of the other. Two variants of the reductionist view may be envisaged: the first portrays aesthetic judgment as an aspect of empirical judgment, while the second reduces empirical judgment to aesthetic judgment.

The first variant of reductionism amounts to the claim that whether scientists attribute aesthetic value to a theory is determined entirely by its degree of empirical success. This claim entails that a scientist takes aesthetic pleasure in a theory when he or she recognizes and approves of the theory's empirical properties, such as its accord with a body of data and its internal consistency. This variant of reductionism has been defended by J. W. N. Sullivan thus: "Since the primary object of the scientific theory is to express the harmonies which are found to exist in nature, we see at once that these theories must have an aesthetic value. The measure of the success of a scientific theory is, in fact, a measure of its aesthetic value, since it is a measure of the extent to which it has introduced harmony in what was before chaos."[18] This view has some notable implications. For instance, it entails that aesthetic judgments of theories are valid or invalid objectively: any aesthetic appraisal of a theory either correctly reflects the theory's degree of empirical success and is thus valid,

18. Sullivan (1919), p. 275. Sullivan's views are further discussed by Kivy (1991), pp. 180–185.

or is at variance with it and is thus invalid. In contrast, prevailing accounts of aesthetic judgments regard them as being valid or invalid only relative to beholders.

According to the second variant of reductionism, "empirical success" denotes a property in virtue of which a theory gives aesthetic pleasure, so that an appraisal of the empirical success of a theory is nothing but a manifestation or aspect of an appraisal of the theory's aesthetic merit.[19] The import of this claim depends on whether we define empirical success to be an observer-independent or an observer-relative property. If we define it as an observer-independent property, as is customary, then this variant of reductionism amounts to the claim that whether we attribute aesthetic value to a theory is determined by the degree of empirical success that the theory intrinsically possesses—a claim indistinguishable from the one discussed in the previous paragraph. If we define empirical success to be an observer-relative property, so that a theory may have empirical success for one observer and lack it for another, then this variant of reductionism amounts to the claim that for a theory to have empirical success is merely for it to accord with an observer's aesthetic preferences.

Reducing empirical judgment to aesthetic judgment might appeal to those who hold to representationalism in art theory. According to representationalism, the degree of the accuracy with which an artwork depicts objects is one of the factors that determine its degree of aesthetic merit. This doctrine stands in opposition to formalism, which regards the aesthetic merit of a work as a matter only of its intrinsic or formal properties. Someone who espouses representationalism in art theory would be inclined to say that the gaze that scientists turn onto their theories is primarily aesthetic: whereas scientists may still wish their theories to be true or empirically adequate, a theory's truth or adequacy should be regarded as, at root, an aesthetic attainment.[20]

Each of these variants of reductionism rules out that aesthetic and empirical judgment could supplement one another as sources of information about the empirical success of theories. On the first variant, the properties of a theory that would activate a particular aesthetic judgment would already have been registered in the empirical appraisal of it. On the second variant, according to which empirical judgment is an aspect of aesthetic judgment, all evaluations of the empirical success of theories

19. Zemach (1986) contends that all criteria of theory assessment applied by scientists, including those that we customarily call empirical, are in fact aesthetic.
20. Representationalism is defended in art theory and extended to the aesthetic appreciation of scientific theories by Kivy (1991).

are performed by aesthetic judgments. In either event, we cannot look to aesthetic judgment to augment our capacity to discern high degrees of empirical adequacy in theories.

Both variants of reductionism face adverse empirical evidence, consisting in the fact that scientists appear ready to pass aesthetic judgments of theories that are at odds with their empirical judgments of them. Examples will be found throughout this book, but a sampling can be given here. Erwin Schrödinger has a high aesthetic opinion of Jean-Baptiste Lamarck's theory of evolution, but he does not thereby claim that it is close to the truth: he writes that it is "beautiful, elating, encouraging and invigorating," but adds, "Unhappily Lamarckism is untenable. The fundamental assumption on which it rests, namely, that acquired properties can be inherited, is wrong."[21] Commenting on the steady-state cosmological theory of Fred Hoyle and others, Denis Sciama showed the independence of his aesthetic and empirical judgments: "It is very beautiful but it is now in serious conflict with observation."[22] Even Einstein seemed to admit at times that beauty in a theory does not secure its truth: in the 1920s he described Arthur S. Eddington's field theory as "beautiful but physically meaningless," and his own attempted unification of the theories of gravitation and electromagnetism as "very beautiful but dubious."[23] These scientists are demonstrating readiness to give a theory a high score in an aesthetic evaluation but a low score in an empirical assessment. This suggests that scientists' aesthetic and empirical judgments are not manifestations or aspects of one another as reductionism maintains.

In the course of this chapter, I have tried to show that an adequate account of scientists' aesthetic and empirical judgments is offered neither by the view that these judgments are entirely uncorrelated with one another nor by the view that they are reducible to one another. Aesthetic and empirical judgments are interrelated in more complex ways, for which we need a more sophisticated model than either autonomism or reductionism.

21. Schrödinger (1958), pp. 21–22.
22. Quoted from Osborne (1986a), p. 12. A similar phrase is attributed to Sciama by Kippenhahn (1984), p. 153: "The steady-state theory has a sweep and beauty that for some unaccountable reason the architect of the universe appears to have overlooked." The import of these statements is identical: the steady-state theory's failure to be instantiated in the universe is revealed by a predictive inadequacy which is quite independent of its aesthetic virtue. The steady-state cosmological theory is put forward in, for example, Hoyle (1950), pp. 108–113.
23. These remarks by Einstein are quoted from Kragh (1990), p. 287.

Two Erroneous Views of Scientists' Aesthetic Judgments

The Inductive Construction
of Aesthetic Preference

1. PRECEPTS AND THEIR WARRANTS

A precept is an instruction directing that a particular methodology or policy be applied, for example, "Formulate bold hypotheses and strive thereafter to refute them." The warrant of a precept is a justification of it, or a reason for believing that it is advisable to obey it. Whether a precept is warranted is determined by two things: the goals of the agent to whom the precept applies and the situation in which the agent operates. A precept is warranted for an agent if obeying the precept contributes to achieving the agent's goals in the agent's situation.

Whether a given precept is warranted is for agents to discover: indeed, one of the problems of action in every field is to ascertain which precepts are warranted for a given goal and situation. There are two ways to discover whether a precept is warranted: I call these goal analysis and inductive projection.[1]

The precepts that goal analysis reveals to be warranted are those that can be inferred by ends–means reasoning from a statement of a set goal. For any goal, there are necessary and sufficient conditions for its achievement, and some of these conditions can be inferred from a statement of the goal. Therefore, once a goal has been set, we may ascertain whether a precept is warranted by investigating, through analysis of the goal, which actions would be likely to promote its achievement. The precepts directing that these actions be performed are the warranted precepts:

1. A previous discussion of the warrants of precepts in science, with parts of which my own account agrees, is Laudan (1984), pp. 23–41.

they are warranted because obeying them will promote the goal's achievement.[2]

There are many alternative goals that may be stipulated for science: the goals of accounting for observational data, of identifying the entities and mechanisms that underlie phenomena, of maximizing our power over our environment, and so on. Once we have attributed one such goal to science, we may analyze what is logically presupposed in the idea of fulfilling it. This analysis can reveal which policies are apt to promote attainment of the goal, or stand to the goal as means to ends. The precepts requiring that these policies be followed are those that goal analysis shows to be warranted in this case.

For example, let us envisage attributing to science the goal of accounting in observational terms for everyday experience. Fulfillment of this goal will be promoted most effectively by a policy of formulating phenomenological generalizations of moderate precision and low scope. Goal analysis shows that precepts calling for this policy to be followed are warranted in this case. In comparison, policies consisting of investigating effects that occur only under exceptional conditions, postulating unobservable entities, and striving for theoretical unification would be much less suited to fulfilling the goal mentioned: the precepts recommending these policies would not be warranted in this case.

Many philosophers have employed or suggested employing goal analysis to identify which precepts are warranted in science. One of these was Descartes. He attributed to natural philosophy the goal of constructing knowledge about the external world that is immune to skeptical doubt. By considering what is implicit in the notion of indubitable knowledge about the external world, and by appeal to particular beliefs about God, space, and material bodies, Descartes concluded that the sole manner for natural philosophers to attain this goal was to study physical phenomena in terms of geometrical kinematics, postulate alternative hidden mechanisms to explain appearances, and appeal to empirical data where necessary to decide which of these mechanisms obtain. Goal analysis shows that precepts recommending these policies are warranted, given Descartes's formulation of the goal of natural philosophy and his other assumptions. In the twentieth century, Karl R. Popper used goal analysis to justify the precepts of falsificationism. In Popper's opinion, because of the logical asymmetry between confirmation and refutation, the goal of science cannot be the accumulation of well-confirmed assertions: it can only be the elimination of error. From an analysis of what

2. Theories of ends–means reasoning, also known as practical reasoning and deliberation, are surveyed by Aune (1977), pp. 112–197.

The Inductive Construction of Aesthetic Preference

would be involved in fulfilling this goal effectively, Popper derives the precepts of striving to propose conjectures that are as bold as possible and of seeking implacably to refute them.[3]

Philosophers have invoked goal analysis not only to ascertain which methods scientists should follow but also to explain why they follow the methods that they do. Some writers have suggested that, as a matter of fact, scientific communities use goal analysis to decide which precepts to obey. This suggestion has found favor especially with those who believe that all scientists throughout history have followed the same methods. After all, if every scientist has shared the same goal and has correctly inferred from it which methods are apt to achieve it, there is no reason to expect scientific practice ever to have altered.[4]

Now for inductive projection. The way in which inductive projection shows a precept to be warranted is as follows. Once a goal for an enterprise has been set, there will at any time t exist a strategy which by that time has proved the most effective in promoting the goal's fulfillment. Inductive projection regards as warranted at time t the precept directing that this strategy be implemented. The precept picked out in this way is regarded as warranted on the grounds that it is, at time t, the most effective that has yet been found in promoting fulfillment of the set goal.

For us to recognize by inductive projection which precepts are warranted, we must have means to compare the effectiveness of alternative strategies in fulfilling the goal that we have set. Once we have concluded by inductive projection that a certain precept is warranted, we will doubtless wish to implement it in preference to any alternative precepts that we may have tested. Whenever a new precept is proposed, we may wish to compare the track record of the policy that it recommends with that of the precept that we regard at present as warranted; and we may discover that the new precept is more strongly warranted than the old one. Inductive projection may of course attribute warrants not only to single-policy precepts but also to precepts calling for mixed strategies or combinations of policies.

In the case of many of the precepts that scientists obey, our belief that they are warranted is based on inductive projection. Consider some of the procedures used in clinical trials of drugs. These include the use of controls, in which the responses of the experimental subjects are compared to those of a control group; the use of placebos, in which an inac-

---

3. Popper (1972), pp. 191–205, explicitly infers scientific precepts from considerations on the aims of science.

4. Those who believe that scientists through the centuries use the same method include Scheffler (1967), pp. 9–10.

tive substance is administered to the control group; single and double blinding, in which the identity of the substances administered is concealed from the subjects and the researchers; randomization, in which subjects are allocated randomly to alternative treatments; and crossovers, in which subjects are switched from one treatment to another.[5] Pharmacologists regard the precepts directing that these procedures be followed as more strongly warranted than precepts recommending alternative procedures. On what do they base this belief? Not goal analysis: it is not feasible to derive from a statement of the goal of science precepts so detailed as to require double blinding in drug trials. Rather, the belief that these precepts are warranted is based on inductive projection. Pharmacologists have found empirically that achievement of their goals is better promoted by obeying these precepts than by obeying others. Because procedures such as double blinding deliver greater success than alternative procedures, the precepts requiring that these procedures be followed are regarded by inductive projection as warranted.

Several writers have recommended that scientists should use inductive projection to validate their precepts. One of these is Nicholas Rescher: in the "method Darwinism" that he once advocated, scientists regard as warranted the method that performs best in a competition with alternative methods.[6] Other writers have suggested that inductive projection is in fact used by scientific communities to justify precepts. For instance, Larry Laudan writes:

Scientific theories not only inspire new theories of methodology, they also—in a curious sense—serve *to justify those methodologies*. For instance, the success of Newton's physics was thought to sanction Newton's rules of reasoning; Lyell's geological theory was cited as grounds for accepting methodological uniformitarianism; the kinetic theory of gases and Brownian motion were thought to legitimate epistemological realism; these are but a few examples of a very common phenomenon.[7]

Laudan claims that in each of these episodes the success of particular theories was taken to justify scientists' continuing to adhere to the method that yielded those theories. In other words, a method was regarded as warranted if its track record was good, as inductive projection suggests.

Inductive projection is frequently used in practices other than science

5. On procedures in the clinical trials of drugs, their justification, and their history, see Pocock (1983).
6. Rescher (1977), pp. 140–166.
7. Laudan (1981), p. 16.

The Inductive Construction of Aesthetic Preference

to ascertain which policies are warranted. For example, financial institutions use inductive credit approval systems to decide to whom to lend money. Such a system identifies which properties, such as age, salary, value of assets, and so on, have been associated with creditworthiness in a sample of people whose financial history is known. In this situation, the warranted precept is the one directing that credit be offered to applicants showing those properties found to be most strongly correlated with creditworthiness.

Let us compare goal analysis and inductive projection as techniques for ascertaining the warrants of precepts. The difference between warrants detected by goal analysis and those detected by inductive projection emerges most clearly when we consider how they are affected by changes in circumstances. Discovering by goal analysis that a precept is warranted involves deriving the precept from a statement of the goal attributed to an enterprise. Provided that it is logically valid, this derivation establishes conclusively that the precept is warranted. Therefore, unless the goal attributed to science is revised or an earlier analysis of that goal is found faulty, a scientist would never have reason to abandon a precept to which a warrant had been attributed by goal analysis. In particular, contingent events cannot provide grounds to suspend or override precepts whose warrant has been revealed by goal analysis. For instance, if Popper's analysis of the goal of science were correct, falsificationism would be the sole method that scientists would ever be justified in following, irrespective of future changes in circumstances.

Warrants revealed by inductive projection do not have the same immutability. However long a precept to which a warrant has been attributed by inductive projection remains more effective than alternative precepts in furthering a set goal, one cannot rule out that in changed circumstances it might prove less effective. For an illustration, consider the procedures followed in drug trials. The reason double blinding and randomization are more effective than alternative procedures in furthering the goals of pharmacologists lies in the characteristics of humans. For example, the warrant of double blinding derives from the fact that the responses of subjects and experimenters in drug trials are altered by knowledge of which substances have been administered. But we cannot rule out that if the characteristics of humans changed in particular ways, double blinding and randomization might become less effective in furthering the goals of pharmacologists than alternative procedures. By inductive projection, we would then attribute a warrant to the precepts recommending these other procedures. Similarly, however long methodological uniformitarianism has yielded success in geology, we cannot rule out that in changed circumstances the negation of uniformitarianism may prove more successful. Inductive projection will then lead us to regard an alternative to uniformitarianism as warranted.

In summary, contingent events can compel us to withdraw the warrants that we attribute to precepts by inductive projection, whereas the attribution of warrants to precepts by goal analysis is conclusive. This dissimilarity is a manifestation of the familiar asymmetries between assertions that are established a priori and assertions established inductively.

Furthermore, the warrants that are attributed to precepts by inductive projection may be weak. What determines the strength of the warrant that a precept is accorded by inductive projection is the precept's degree of effectiveness in fulfilling the goal set to an enterprise. If a precept invariably yields success, inductive projection will show its warrant to be strong. If a precept yields success only rarely, its warrant may be weak—though it may be far from negligible, since there is compelling reason to follow a strategy that has a low probability of success if there is no better-performing strategy available. In contrast, all warrants discovered by goal analysis are strong. After all, if it is intrinsic to the notion of the goal of an enterprise that in order to achieve that goal it is necessary to perform a certain act, then a precept calling for this act to be performed is strongly warranted.

But inductive projection has some advantages as a technique for revealing the warrants of precepts. Thanks to inductive projection, we can recognize a precept as warranted even if we are incapable of deriving it by ends–means reasoning from a statement of the goal of an enterprise. If we are to recognize that a precept is warranted by goal analysis, its functionality for the achievement of a goal must be clear to us from analysis of the goal: in contrast, inductive projection allows us to recognize the warrant of precepts whose success we would not have been able to predict from first principles.

2. The Warrant of Empirical Criteria

In the previous section, I described goal analysis and inductive projection as techniques for revealing the warrants of precepts. I now restrict the discussion in two respects. First, I wish to focus on the use of goal analysis and inductive projection to reveal the warrants of precepts that consist of criteria for the evaluation of scientific theories. Second, I wish to ascertain the place of goal analysis and inductive projection in actual scientific practice. My intention is therefore to use goal analysis and in-

The Inductive Construction of Aesthetic Preference

ductive projection to model the reasoning by which scientists come to lend weight to the criteria for theory evaluation that they in fact apply.[8]

First, a logical point. It is logically impossible to rely on only inductive projection to ascertain which criteria for theory evaluation are warranted. Here is the reason. In order to discover by inductive projection that a particular criterion is warranted, we must examine some theories that are good (under any conception of goodness that we may possess) to determine which other properties they also possess. If we discover a correlation between theories' being good and their exhibiting some other property P, we will conclude that the criterion "Other things being equal, prefer a theory that shows P to one that does not" is warranted. Now, the first step of this procedure requires us to recognize theories that are good, or that fulfill to some extent the goal that we set to theory evaluation. Recognizing theories that stand in this relation to our goal requires a criterion for theory evaluation of which we can say by goal analysis that it is warranted. Therefore, any attempt to discover which criteria for theory evaluation are warranted must begin by applying goal analysis to the goal that is set to science: only thereafter may inductive projection be applied to discern which further criteria are warranted.

We may therefore expect modern scientific communities to formulate some criteria for theory evaluation which, given the goal that they prescribe to themselves, goal analysis shows to be warranted. I suggest that the criteria for theory evaluation that satisfy this description are what I called in Chapter 1 scientists' empirical criteria. I assume, as does the logico-empirical model of theory assessment, that modern scientific communities attribute to science the goal of formulating theories that possess the highest possible degree of empirical adequacy, or truth. To be able to recognize theories that contribute to fulfilling this goal, a scientific community must identify some properties of theories that are indicative of high degrees of empirical adequacy. These properties are identified by analyzing what sort of attainment for a theory is a high degree of empirical adequacy and what other properties a theory has to possess to achieve it. An analysis of this sort will conclude that for a theory to have a high degree of empirical adequacy it must possess such properties as consistency with extant empirical data and with current well-corroborated theories, novel prediction, explanatory power, empirical content, and internal consistency. The criteria for theory evaluation that express

8. In an approach similar to mine, Newton-Smith (1981), pp. 224–225, suggests that our belief in the warrant of some criteria for theory evaluation is based on inductive projection; Watkins (1984), pp. 166–224, proposes a set of criteria derived by goal analysis from the "optimum aim" that he attributes to science.

preference for theories showing these properties are the empirical criteria that I listed in Chapter 1.

To recognize these criteria for theory evaluation as warranted, no induction over the performance of theories is required. Our reason for wanting theories to show internal consistency or consistency with extant empirical data is not that we have detected these properties in many previous theories with high degrees of empirical adequacy and expect this correlation to hold in future theories. Rather, it is that we find ourselves referring to these properties in explicating what it is for a theory to have a high degree of empirical adequacy. Indeed, it is nonsensical to set out to uncover a correlation between high degrees of empirical adequacy on the one hand and internal consistency or consistency with extant empirical data on the other: we would be unable even to specify the meaning of "empirical adequacy" without using the latter notions. These considerations are grounds for concluding that scientists' empirical criteria acquire their warrant not by inductive projection but by goal analysis.

3. The Aesthetic Induction

Once a scientific community has formulated by goal analysis a set of criteria for theory evaluation that are warranted in the light of the goal that it has attributed to science, it will wish to consider whether it is able by inductive projection to discover any further criteria that are warranted in the light of this goal. After all, the community cannot be confident that its analysis of the goal has been so perceptive as to identify every precept that will promote its fulfillment. The community would benefit from knowing if, for example, a strong correlation exists between a theory's satisfying the criteria formulated by goal analysis and its showing some further property P: discovering such a correlation would allow the community to formulate an additional criterion for theory evaluation, "Prefer theories that show P," as an extra diagnostic tool to identify theories that fulfill the criteria formulated by goal analysis.

I have suggested that, by goal analysis, modern scientific communities identify properties of theories that promote the attainment of high degrees of empirical adequacy: these are the empirical properties of theories. Do modern scientific communities also employ inductive projection to ascertain if there are other properties of theories that are correlated with the empirical properties? I claim that they do, and I claim that among the properties of theories that they examine for these correlations are aesthetic properties.

Consider two of the striking features of scientists' aesthetic evaluations of theories which we uncovered in Chapter 3. First, the degree of favor with which any aesthetic property of theories is regarded by scientific communities varies with time. Every property that has at some date been seen as aesthetically attractive in theories has at other times been judged displeasing or aesthetically neutral. Second, the degree of favor with which scientists have regarded an aesthetic property appears to have responded to the empirical performance of theories that possess that property. If a theory possessing an aesthetic property P scores notable empirical success, the community comes to regard P with increased favor and to expect future theories showing P to be successful too. On the other hand, if there later arise theories that lack P but are empirically more successful than the P-bearing theories, then the community's preference for future theories to show P wanes. An example of these responses is given by the succession of models invoked in physiology and neurophysiology.

These features suggest that, in passing aesthetic evaluations of theories, scientists are conducting a search by inductive projection for properties of theories that are correlated with high degrees of empirical success but that are not already valued by scientists' empirical criteria for theory evaluation. Let us give this suggestion a more formal expression.

As described in Chapter 2, we imagine a scientist's or community's aesthetic canon to consist of innumerably many criteria. Each criterion names a property of theories and specifies a weighting that indicates the degree of value attached to that property in theory evaluation. In the canon of any actual scientist or community, the weightings of most of the properties will be zero, indicating that no aesthetic value is attached.

I propose the following model of the mechanism by which scientific communities formulate their aesthetic canons for theory evaluation. A community compiles its aesthetic canon at a certain date by attaching to each property a weighting proportional to the degree of empirical adequacy then attributed to the set of current and recent theories that have exhibited that property. The degree of empirical adequacy of a theory is, of course, judged by applying the community's empirical criteria for theory evaluation. I name this procedure the aesthetic induction.[9]

Here is an illustration of the working of the aesthetic induction. A scientific community looks back over the recent history of a particular branch of science. It perceives that some theories, which are to a notable degree visualizing (rather than abstract) theories, have been empirically very successful, whereas others, which lend themselves to mechanistic

9. I first proposed an inductive mechanism for the origin of scientists' aesthetic canons for theory evaluation in McAllister (1989), pp. 36–41.

analogies, have won little empirical success. Both visualization and tractability by mechanistic analogies are aesthetic properties of theories. In consequence of the empirical success of the visualizing theories, the property of visualization will obtain an increased weighting in the aesthetic canon for theory evaluation that the community will hereafter apply. By contrast, the property of being tractable by mechanistic analogies will receive a lowered weighting in the canon, in virtue of the scarce empirical success of recent theories that displayed this property.

The aesthetic induction is an instance of inductive projection, since it amounts to consulting the properties of past good theories to determine which future theories should be expected to be good. A particular theory's achieving notable empirical successes at some date contributes to determining which theories will later be embraced by the community. What is peculiar to this instance of inductive projection is that the properties that are examined, both in past theories (whose empirical performance is documented) and in future ones (whose performance is unknown), are aesthetic properties of theories.

The expectation of scientists that theories sharing the aesthetic properties of past empirically successful theories will reproduce their success is probably never expressed explicitly. Indeed, for the most part, scientists' aesthetic preferences among theories are acquired and applied unselfconsciously. But these facts do not count against my suggestion that scientists' aesthetic preferences are formed by inductive projection: we frequently perform inductive projections unselfconsciously, both within and outside science.

By imagining the aesthetic induction in operation, we can infer how a community's set of aesthetic preferences among theories will evolve in particular circumstances. A theory that achieves significant empirical success will cause its community's aesthetic canon to be remodeled to a certain extent, in such a way that the canon comes to attribute a greater weighting to that theory's aesthetic properties. The aesthetic canon will therefore acquire a bias in favor of any future theories that exhibit the aesthetic properties of current successful theories. In other words, by their empirical success, theories can predispose the community to choosing future theories with properties similar to their own. A future theory will then win endorsement from the aesthetic canon in the measure to which it shares the aesthetic properties of current theories that have been attributed high degrees of empirical adequacy. If, on the other hand, a new theory shows aesthetic properties different from those currently entrenched in the canon, it will be denied endorsement by the aesthetic canon.

However, the degree of empirical adequacy that the community attributes to the theories that it has embraced tends in the long term to decrease with the passage of time, because of the discovery of empirical data unfavorable to them. (This remark is merely a particular expression of the general principle that any theory with empirical content will someday be found to be in error.) If the degree of empirical adequacy attributed to some theory T decreases, inductive projection has the effect that the weighting of its aesthetic properties within the canon will also decrease. If new theories possessing those same aesthetic properties arise at such a time, they will not win as much favor as they would have if they had appeared earlier, when their aesthetic properties enjoyed greater weightings. On the other hand, if at such a time theories arise that have new aesthetic properties, they will encounter less resistance from the community's aesthetic canon than they would have earlier.

The intensities with which degrees of empirical adequacy affect the weightings of aesthetic properties within a canon may in various ways be filtered. For instance, a community may resolve to attach greater importance to the empirical performance that theories have shown in the recent past than to their performance decades ago, or to empirical successes gained in fields of active research than in fields that are currently stagnant. Such preferences amplify or diminish the effect that a particular theory's empirical success has on the weightings of aesthetic properties in the canon.

My view that scientists come to attach aesthetic value to a theory according to the degree of empirical success scored by it and by theories that share its aesthetic properties is not quite an expression of aesthetic functionalism. According to functionalism, the principal factor determining the degree to which an object is regarded as beautiful is its fitness to a purpose. This doctrine was expressed for instance by William Hogarth in the following terms:

> Fitness of the parts to the design for which every individual thing is form'd, either by art or nature, is [. . .] of the greatest consequence to the beauty of the whole. This is so evident, that even the sense of seeing, the great inlet of beauty, is itself so strongly bias'd by it, that if the mind, on account of this kind of value in a form, esteem it beautiful, tho' on all other considerations it be not so; the eye grows insensible of its want of beauty, and even begins to be pleas'd, especially after it has been a considerable time acquainted with it.[10]

I endorse the functionalist conviction that fitness to a purpose is an important factor determining aesthetic valuations. However, the aesthetic

10. Hogarth (1753), p. 32.

induction does not consist in attributing beauty to each scientific theory to the degree to which it shows fitness to its purpose. On the contrary, the aesthetic induction will cause scientists to regard as beautiful a theory that shows little fitness to its purpose—i.e., empirical success—if this theory shares the aesthetic properties of theories that scored notable empirical success.

My claim that scientists attach aesthetic value to properties of theories associated with empirical success has some affinity with evolutionary accounts of aesthetic judgment. According to some versions of evolutionary theory, the members of certain bird and mammal species conduct mate choice partly on aesthetic preferences acquired by natural selection. These aesthetic preferences are biologically advantageous since they predispose organisms to choose as a mate an organism that shows features associated with good reproductive potential, such as bright plumage.[11] Sociobiology claims that, similarly, the human propensity to invest some objects with aesthetic value has been acquired by natural selection.[12] One of the ways in which such a propensity might be biologically advantageous is by causing humans to have pleasure reactions to features of objects that they have learned to associate with utilitarian value. The aesthetic induction in science might be regarded as an instance of this predisposition. As always, a sociobiological account must be qualified by an acknowledgment that human behavior cannot be explained fully without extensive reference to cultural factors.

On my model, the degree of overall favor won by a theory is the resultant of the community's evaluations of it on empirical and aesthetic criteria. Therefore, part of the perceived merit of a theory comes to it from its current empirical performance: this is the component assessed on the application of empirical criteria. The other part of its perceived merit derives from the record of empirical performance that it and theories with similar aesthetic properties have built up in the past: this record of performance is mediated through the process of inductive projection. Let us now take a closer look at the relation between scientists' aesthetic and empirical evaluations of their theories.

4. The Conservatism of Aesthetic Canons

In what follows, P represents one of the possible aesthetic properties of theories, such as a form of symmetry or a form of analogical tractability.

---

11. Some views of evolutionary biologists on the role of aesthetic criteria in sexual selection are described by Cronin (1991), pp. 183–204.

12. For an introduction to sociobiological accounts of aesthetic judgment, see Lumsden (1991).

E_P is the degree of empirical adequacy that a community attributes to the set of its theories that possess P. The magnitude of E_P varies with time, since both the acquisition of new empirical data and the formulation of new P-bearing theories affect the degree of empirical adequacy attributed to the set of theories that have P. Lastly, as in Chapter 2, W_P represents the weighting given to property P in the community's aesthetic canon for theory evaluation. The magnitude of W_P varies in response to changes in E_P. The aesthetic induction, by which the community fixes the weightings that at some instant characterize its aesthetic canon, amounts to computing the magnitude of W_P from the magnitude of E_P at that instant, as well as the weightings of all other properties to which the canon refers.

Since E_P is the degree of empirical adequacy that a community attributes to the set of all its P-bearing theories, the empirical performance of any further P-bearing theory embraced by the community typically has only a small effect on the magnitude of E_P. For example, if the set of P-bearing theories has a poor empirical record, and therefore E_P is low, the success of a prediction by a new P-bearing theory will raise E_P by only a small amount. Similarly, if the empirical record of the set of P-bearing theories is good, and therefore E_P is high, an empirical failure by the latest such theory to be embraced will depress E_P only slightly. Since the magnitude of W_P is determined by the magnitude of E_P rather than by the empirical performance of the latest P-bearing theory, this means that the aesthetic canon shows a damped response to changes in the empirical performance of the set of P-bearing theories.

The fact that aesthetic canons do not at all times accord with the empirical performance of the latest theories may be viewed as a particular instance of the fact that, in a changing environment, an evolving system is unlikely to be able to readjust all its properties so that they are always optimal for the prevailing circumstances. At any time an evolving system is likely to show atavisms, or features that, while they may have been optimal for earlier circumstances, are suboptimal now. For example, human cultures invariably contain both elements for which there is a current justification on utilitarian grounds and elements that can be justified only by appeal to tradition or heritage.

Because aesthetic canons do not reflect instantly the changing empirical performance of theories, evaluations of theories conducted on aesthetic criteria will tend to lag behind evaluations conducted on empirical criteria. What I mean is, more precisely, the following. Imagine mapping scientific theories onto "aesthetic space," a multidimensional space of which each location represents a particular set of aesthetic properties that theories may possess. A theory is represented in aesthetic space by the point corresponding to the combination of aesthetic properties that it possesses. If the axes are defined suitably, the points representing two aesthetically similar theories adjoin one another. Now imagine representing in this space the sequence of theories that a community adopts in a branch of science. The point representing the community's current theory jumps from one spot to another according to the aesthetic properties of the theories that the community successively adopts. If scientists had no aesthetic preferences in theory choice, the motion of the point would exhibit no systematic pattern; but any aesthetic preference to which the community holds affects the likelihood of the spot's occupying particular regions of aesthetic space, and thus its trajectory in the space.

Now, the aesthetic canon attaches the greatest weightings to the aesthetic properties of the empirically successful theories previously adopted by the community. This ensures that at any time the aesthetic canon tends to prefer theories which in aesthetic space lie in the neighborhood of the community's previous theories. For instance, if a community's empirical criteria recommend switching from a theory T_1 to an aesthetically dissimilar theory T_2, the aesthetic canon would express a preference for theories in the neighborhood of T_1. More generally, if a community's empirical criteria recommend a succession of theories that show a drift in a particular direction, its aesthetic canon tends to resist that drift. In other words, the aesthetic judgments that a community passes on theories will typically be conservative by comparison with its own empirical judgments. In consequence, a theory with a long track record of empirical success is likely to be seen as aesthetically pleasing, since its aesthetic properties will have acquired great weighting in its community's aesthetic canon: an aesthetically innovative theory is likely to be seen as perceptually unappealing for a period following its formulation, since its aesthetic properties neither conform to the canon that they found in existence nor have yet altered the canon in their favor.

Thomas Henry Huxley expresses some of the effects of the aesthetic induction in his aphorism about "the great tragedy of Science—the slaying of a beautiful hypothesis by an ugly fact—which is so constantly being enacted under the eyes of philosophers."[13] The beauty of a hypothesis derives from an accord between its aesthetic properties and the community's aesthetic canon. We may take Huxley's "fact" to be a low-level theory. If a fact has been uncovered only recently, the theory that expresses it may have aesthetic properties which are unfamiliar to the community, and which therefore have as yet low weightings in the aesthetic canon. This is why a fact appears ugly. Nevertheless, empirical criteria

13. Huxley (1894), p. 244.

The Inductive Construction of Aesthetic Preference

may recommend embracing the ugly fact in preference to the beautiful hypothesis. Huxley might draw consolation from the following thought: by continued application of the aesthetic induction, the community will come gradually to see the fact as beautiful.

If few writers devote attention to the role of aesthetic criteria in the assessment of scientific theories, the number of those who note the time lag of aesthetic appreciation behind empirical judgment is tiny. Jeremy Bernstein writes: "In science as in the arts, sound aesthetic judgments are usually arrived at only in retrospect. A really new art form or scientific idea is almost certain at first to appear ugly. The obviously beautiful, in both science and the arts, is more often than not an extension of the familiar. It is sometimes only with the passage of time that a really new idea begins to seem beautiful."[14] My model of the aesthetic induction offers a mechanism to explain what for Bernstein is a phenomenological observation, that an aesthetically novel, empirically successful theory will develop in the beholder's eye from ugly duckling to beautiful swan. Similarly, Penrose writes: "Perhaps one's aesthetic judgments will change [. . .]. Such judgments are, in any case, often to a considerable extent, acquired tastes. In these cases one cannot really appreciate the beauty of something until some familiarity with it has been gained—one has really to have thought about it for quite a while."[15] Of course, to come to see a new theory as beautiful, it is not sufficient just to mull it over: it is necessary for the community to alter its aesthetic canon in response to the empirical success of the new theory and in accord with the new theory's aesthetic properties.

The time lag of aesthetic appreciation behind empirical judgment manifests itself not only in scientists' preferences among theories but also in our preferences among representations of theories, for example in the form of texts. According to Bertrand Russell, readers are more disposed to attribute elegance to a statement of orthodox ideas than to a statement of novel ideas, since the literary style that each epoch regards as graceful is best suited to expressing that epoch's orthodoxy:

> Broadly speaking, old conceptions have acquired pleasant literary clothes, whereas new ones still appear uncouth. An aesthetic bias in favour of good literary form is therefore likely to be associated with conservatism. [. . .] As a result of many centuries of Platonism, the language of educated men can now express even the most difficult of Plato's ideas without crabbiness; but this was not the case in his own day. [. . .] In

14. Bernstein (1979), p. 3.
15. Penrose (1974), p. 267.

such ways those who insist upon elegant literary form are compelled to lag behind—often far behind—the best thought of their time. Per contra, conservatives have a great aesthetic advantage over innovators, for ideas [. . .] grow more beautiful as they grow older.[16]

The parallel between the aesthetic induction operating in the assessment of theories and the quasi-inductive mechanism for the evolution of literary styles that Russell identifies does not, of course, nullify the distinction between scientific theories and their representations.

5. Fashions and Styles in Science

The features of the aesthetic induction suggest that scientists' canons for theory evaluation will be prey to fashions, and that sequences of theories adopted in a branch of science will be characterized by aesthetic styles. To say that theorizing is characterized by styles is to claim that periods of theorizing can be identified in which theories show common aesthetic properties, which are absent or infrequent among theories of other periods.

Suppose that a newly formulated theory T, possessing particular aesthetic properties, scores great empirical success. In virtue of the aesthetic induction, the aesthetic properties of T will win an increased weighting in the community's aesthetic canon. In due course, the community will wish to adopt further theories, to account for domains of phenomena other than that of T. The aesthetic canon disposes the community to adopting theories with aesthetic properties similar to those of T. Suppose some such theories are found and tested for their empirical virtues. If none of these theories is empirically successful, the weightings of the aesthetic properties of T in the canon will not be increased. But if these theories have empirical success, this success will further increase the weightings attributed to the aesthetic properties of T. This cycle of events may be repeated, amplifying the weightings each time. So, once the aesthetic properties of some successful theory T have received an initial favorable weighting, the community acquires a disposition to act in such a way as to increase this weighting. Those aesthetic properties have become fashionable.

A fashion is dissipated once theories possessing the aesthetic properties that the fashion prescribes are shown to be empirically less successful than theories that show other aesthetic properties. Observation of this

16. B. Russell (1940), p. 457.

empirical inferiority would depress the weightings of the hitherto fashionable aesthetic properties within the aesthetic canon.

If, in a given period, certain aesthetic properties are in fashion, their incidence among theories formulated or adopted during that period will be higher than among theories formulated or adopted at other times. Because of this, a record of the theories that a community has successively embraced will show periods in each of which a particular set of aesthetic properties is dominant. It is natural, drawing inspiration from history of art, to associate such sets of aesthetic properties with styles.[17] Each style characterizes a sequence of theories put forward or adopted during the history of a science.

For a visualization of the occurrence of styles in theory choice, let us turn once more to aesthetic space, where any given theory is represented by a point corresponding to its particular set of aesthetic properties, and a community's sequence of choices among theories amounts to a series of jumps from one location to another. In a period during which theories are characterized by a given style, the points on which the community alights in aesthetic space are restricted to a particular region, determined by the aesthetic properties that the current fashion prescribes. When this fashion dissipates, the points on which the community alights cease to cluster. The community may next adopt a theory with quite different aesthetic properties, alighting on a point in a distant region of aesthetic space. If the new theory is empirically successful, a new fashion may develop centered on the aesthetic properties that it shows.

The suggestion that scientific theorizing exhibits styles has been put forward before. For instance, I. Bernard Cohen has shown that successive theories in physical science from the end of the seventeenth century onwards exhibited a Newtonian style: they resembled one another in, among other respects, seeking to analyze physical systems as resulting from the effect of radial forces, acting at a distance.[18] Elements of several other styles of scientific work have been identified by Alistair C. Crombie.[19] However, what has so far lacked in discussions of styles in science is a model by which their formation, endurance, and decline were explained. The aesthetic induction constitutes such a model.

6. An Example of Scientific Style: Mechanicism

An instructive example of style of theorizing is mechanicism, which characterized physical theorizing especially in the second half of the

17. Some issues involved in speaking about styles in science are surveyed by Wessely (1991) and Hacking (1992).
18. I. B. Cohen (1980), pp. 52–154.
19. Crombie (1994).

nineteenth century. Mechanicism prescribed that phenomena outside the domain of mechanics—such as optical, thermal, or electromagnetic phenomena—should be represented by physical theory as the effect of mechanical phenomena, such as motions, collisions, and tensions. The formation, endurance, and decline of this style of theorizing are well explained by the aesthetic induction: mechanicism came into fashion on the strength of the accumulated empirical record of mechanistic theories, but declined when it was seen to prevent physicists from embracing theories that were empirically successful.

The strategy of seeking mechanicist accounts of phenomena outside the domain of mechanics was of course inaugurated by Newton, who for instance described light as the propagation of particles. During the eighteenth century, mechanical models were developed of elasticity and of fluid flow; by the end of that century, heat too had come to be regarded as the effect of particle motions. Mechanicist theories in these domains built up an impressive empirical record.

By the mid–nineteenth century, on the strength of this empirical record, the strategy of seeking mechanicist accounts of all phenomena was elevated into an explicit goal of physical theorizing. The goal was expressed most forcefully by William Thomson (later named Lord Kelvin) in the series of lectures that he gave at Baltimore in 1884:

> My object is to show how to make a mechanical model which shall fulfil the conditions required in the physical phenomena that we are considering, whatever they may be. At the time when we are considering the phenomenon of elasticity in solids, I want to show a model of that. At another time, when we have vibrations of light to consider, I want to show a model of the action exhibited in that phenomenon. [. . .] It seems to me that the test of "Do we or not understand a particular subject in physics?" is, "Can we make a mechanical model of it?"[20]

Under this program, theorizing in electromagnetism involved devising arrangements of rods, wheels, weights, and springs that would replicate the behavior of molecules and the electromagnetic ether. For example, in 1862 Maxwell developed a model of the ether involving wheels and vortices that accounted for several aspects of the propagation of light and its behavior in magnetic fields. Similarly, Thomson devoted much effort to elucidating the microscopic structure that the ether would have to possess in order to be both elastic enough to transmit the transverse waves of light and yet not so viscous as to resist the motion of bodies. Ludwig

20. W. Thomson (1884), p. 111; for a similar formulation see p. 206.

The Inductive Construction of Aesthetic Preference

Boltzmann was equally keen to find mechanicist models of phenomena. Moreover, he appears to have drawn aesthetic satisfaction from such devices, as Paul Ehrenfest, who studied with Boltzmann, describes:

> He obviously derived intense aesthetic pleasure from letting his imagination play over a confusion of interrelated motions, forces and reactions until the point was reached where they could actually be grasped. This can be recognized at many points in his lectures on mechanics, on the theory of gases, and especially on electromagnetism. In lectures and seminars Boltzmann was never satisfied with just a purely schematic or analytical characterization of a mechanical model. Its structure and its motion were always pursued to the last detail. If, for example, several strings were used to illustrate certain kinematical relations, then the conceptual arrangement had to be devised in such a way that the strings would not become entangled.[21]

As the range of known electromagnetic phenomena grew, however, physicists found that mechanical models to account for them were becoming increasingly intricate. The complexity of some of Thomson's and Boltzmann's models reached baroque dimensions, and the burden imposed on theorizing by the requirement of providing mechanicist interpretations of phenomena became stifling.[22] Some physicists started to construct theories that regarded electromagnetic parameters as primitive entities not requiring interpretation in mechanical terms. For example, Maxwell eventually abandoned the search for mechanicist models of the ether, limiting himself to giving mathematical descriptions of electromagnetic phenomena; similarly, H. A. Lorentz in 1892 developed a theory of the electron that did not depend on treating the electromagnetic field as a mechanical system. Einstein, who was a student at this time, later recalled the period as follows: "One got used to operating with these fields as independent substances without finding it necessary to give oneself an account of their mechanical nature; thus mechanics as the basis of physics was being abandoned, almost unnoticeably, because its adaptability to the facts presented itself finally as hopeless."[23]

The life cycle of mechanism testifies both to the influence that styles have on theorizing and to the intensity of scientists' empirical concerns.

21. Quoted from Klein (1972), p. 72. For further discussion of Boltzmann's style of theorizing, see D'Agostino (1990).

22. Thomson's and Boltzmann's model making is described as baroque by Klein (1972), p. 73. Mechanicism was criticized for the burden that it imposed on theorizing by Duhem (1906), pp. 69–104.

23. Einstein (1949), p. 25.

The mechanicist style of theorizing became increasingly deeply entrenched as long as the theories that it spawned continued to score empirical success; but once mechanism appeared to have begun to hamper the pursuit of empirical success, the majority of physicists abandoned it.

The Inductive Construction of Aesthetic Preference

The Relation of Beauty to Truth

I once heard Dirac say in a lecture, to an audience which largely consisted of students, that students of physics shouldn't worry too much about what the equations of physics mean, but only about the beauty of the equations. The faculty members present groaned at the prospect of all our students setting out to imitate Dirac.

—Steven Weinberg, "Towards the Final Laws of Physics"

1. Beauty as an Attribute of Truth

One of the most remarkable features of modern science is the conviction of many scientists that their aesthetic sense can lead them to the truth. In this chapter, we review some statements of this conviction; we consider how it coheres with other methodological tenets of scientists; and we inquire, in the light of our earlier findings, to what extent this conviction could ever be justified.

Many twentieth-century scientists appear to believe that a beautiful scientific theory is bound to be close to the truth. Heisenberg recalls putting the following thesis to Einstein in 1926: "If nature leads us to mathematical forms of great simplicity and beauty—by forms I am referring to coherent systems of hypotheses, axioms, etc.—[. . .] we cannot help thinking that they are 'true,' that they reveal a genuine feature of nature."[1] In his work on general relativity theory, Penrose formed the impression that there is a correlation between theories' possessing certain aesthetic properties and their being close to the truth:

It is a mysterious thing in fact how something which looks attractive may have a better chance of being true than something which looks ugly.

---

1. Heisenberg (1971), p. 68.

[. . .] I have noticed on many occasions in my own work where there might, for example, be two guesses that could be made as to the solution of a problem and in the first case I would think how nice it would be if it were true; whereas in the second case I would not care very much about the result even if it were true. So often, in fact, it turns out that the more attractive possibility is the true one.[2]

Dirac judges that there are aesthetic grounds for believing that the general theory of relativity is basically true, regardless of the degree of its accord with particular experimental data: "One has a great confidence in the theory arising from its great beauty, quite independent of its detailed successes. [. . .] One has an overpowering belief that its foundations must be correct quite independent of its agreement with observation."[3] Other physicists who seem to have believed that beautiful theories are bound to be true or close to the truth are Minkowski and Hermann Weyl.[4] Aesthetic factors are said to have played a role in convincing some scientists that James D. Watson's and Francis Crick's theory of the structure of DNA is correct: Watson writes that Rosalind Franklin "accepted the fact that the structure was too pretty not to be true."[5]

Some further writers, while showing less confidence that aesthetic criteria are reliable detectors of true theories, find it apt that beauty should accompany truth. For example, Oliver Lodge seems to have found Thomson's vortex theory of the atom aesthetically so pleasing as to feel that it would be appropriate for it to be true. He wrote about this theory: "It is not yet proved to be true, but is it not highly beautiful? a theory about which one may almost dare to say that it deserves to be true."[6] The acceptance by Lodge that a theory could be so beautiful and yet not true marks his distance from Dirac and the other scientists cited above.

The thesis that aesthetic virtues accompany truth, sometimes expressed in the motto *Pulchritudo splendor veritatis* ("Beauty is the splendor of truth"), has a long and glorious genealogy. It is an expression

2. Penrose (1974), p. 267.
3. Dirac (1980a), p. 44. For previous discussions of Dirac's belief that beauty accompanies truth, see Kragh (1990), pp. 275–292, McAllister (1990), and Hovis and Kragh (1993). Further remarks about the incidence of aesthetic considerations in Dirac's judgments of theories are contained in Mehra (1972), especially p. 59. The possibility that beauty may be a guide to truth is aired further in Davies (1992), pp. 175–177.
4. On Minkowski's belief that beauty accompanies truth, see Galison (1979); on Weyl's, see Kragh (1990), p. 287. Tonietti (1985), p. 8, reports, however, that on one occasion Weyl declared: "My work always tried to unite the truth with the beautiful: but when I had to choose one or the other, I usually chose the beautiful."
5. Watson (1968), p. 124.
6. Lodge (1883), p. 329.

The Relation of Beauty to Truth

of the doctrine that there is necessarily an accord between an entity's perceptual features and its practical qualities. This doctrine was widely accepted in classical Greek thought: it is embodied in the term *kalos kagathos*, which may be rendered as "both good to look at and manifesting goodness in action."[7] The doctrine also has a moral expression, in the thesis that handsomeness or comeliness accompanies moral virtues or spiritual nobility in persons, which is found notably in Homer.

If it were the case that beautiful theories are bound to be true or close to the truth, the beauty of a theory would count as evidence that it is close to the truth, and aesthetic criteria could be used to reveal that it is. Many of the scientists cited above suggest that aesthetic criteria for recognizing truths do exist. For instance, the remark of Heisenberg to Einstein quoted above continues as follows: "You may object that by speaking of simplicity and beauty I am introducing aesthetic criteria of truth, and I frankly admit that I am strongly attracted by the simplicity and beauty of the mathematical schemes with which nature presents us. You must have felt this, too."[8]

The claim that there exist aesthetic criteria for recognizing truths is different from the reductionism about scientists' aesthetic and empirical judgments that we discussed in Chapter 4. Since reductionism regards aesthetic judgment and empirical judgment as aspects of one another, it rules out that aesthetic judgment can yield an estimate of the empirical adequacy of theories that does not rely on empirical judgment. According to the claim that we are examining here, in contrast, the aesthetic judgment of scientists is independent of empirical judgment, but it is capable of recognizing truths and is therefore an additional source of information about the empirical merits of theories. In what circumstances this additional source of information about the empirical merits of theories might be useful to scientists is the question that we shall consider next.

2. Aesthetic Judgment and the Recognition of Truth and Falsity

To judge the degree of empirical adequacy of theories, most scientists are content to use empirical criteria such as internal consistency and consistency with extant empirical data. For instance, if two competing theories entail different predictions and there are reliable empirical data that unambiguously accord with the predictions of one theory and conflict with

7. On the concept of *kalos kagathos*, see Dover (1974), pp. 41–45.
8. Heisenberg (1971), pp. 68–69.

those of the other, most scientists in most cases would be happy to conclude that the former theory is empirically superior to the latter.

But, as scientists are aware, there are cases in which empirical criteria for theory evaluation fail to reveal a theory's true degree of empirical adequacy. These cases may be classified as false negatives, in which empirical data appear to cast doubt on a theory that is true, and false positives, in which data appear to corroborate a theory that is false.

Let us begin with the false negatives. Many high-level scientific theories are of such wide scope that they fail on their own to yield any determinate predictions that can be compared with empirical data. To draw determinate predictions from such a theory, it is necessary to conjoin it with a set of auxiliary hypotheses, which make assumptions about the behavior and boundary conditions of the systems under study. Any experiment that purports to test a high-level theory in fact tests the conjunction of the theory and these auxiliary hypotheses. An unfavorable verdict given by an empirical test of this conjunction shows it to be empirically inadequate, but it does not reveal whether the theory itself is false or only the auxiliary hypotheses are false. This is an illustration of what is known as the Duhem-Quine thesis.[9] The latter case is the false negative.

Since any other empirical evaluation of such a theory would similarly involve auxiliary assumptions and thus be equally indeterminate, it seems that in this situation whether a theory should be accepted or rejected can be determined, if at all, only from a nonempirical evaluation. Accordingly, some scientists have suggested deciding on aesthetic criteria whether a given high-level theory should be rejected on the strength of an unfavorable empirical verdict. In the following passage, Dirac considers what the correct response would have been if an experimental test of the general theory of relativity had given an unfavorable verdict:

Suppose a discrepancy had appeared, well confirmed and substantiated, between the theory and observations. [. . .] Should one then consider the theory wrong? [. . .] I would say that the answer to the last question is emphatically No. [. . .] Anyone who appreciates the fundamental harmony connecting the way Nature runs and general mathematical principles must feel that a theory with the beauty and elegance of Einstein's theory *has* to be substantially correct. If a discrepancy should appear in some application of the theory, it must be caused by some secondary feature relating to this application which has not been ade-

9. Duhem (1906), pp. 180–190; Quine (1953), pp. 37–42.

The Relation of Beauty to Truth

quately taken into account, and not by a failure of the general principles of the theory.[10]

If aesthetic judgment were able to reveal that a theory is true or close to the truth, it would solve the problem of detecting the false negatives obtained in the empirical evaluation of theories.

Let us now turn to the false positives. Since there is a valid argument from any false premises to some true conclusion, it is possible for any scientific theory that is false to make predictions that accord with empirical findings. Therefore, it may be that some theories that have passed all empirical tests to which they have been subjected are nevertheless distant from the truth. If there are no more discriminating empirical tests to which they can be subjected, such theories may be identified only, if at all, by appeal to nonempirical criteria. Which nonempirical criteria can reveal the falsity of an empirically successful theory? Dirac suggests that aesthetic criteria have this power: theories that are aesthetically displeasing are likely to be distant from the truth even if they have a track record of empirical success.

Dirac's view becomes apparent in his comments on quantum electrodynamics, a theory that is the quantum counterpart of classical electrodynamics. By the end of the 1940s, this theory had become one of the empirically most successful scientific theories of all time. The values that quantum electrodynamics predicts for such parameters as the Lamb shift (a splitting of one of the spectral lines of hydrogen) and the magnetic moment of the electron agree with the results of measurement within experimental accuracy, which is of a few parts per million. However, in the course of generating these predictions, quantum electrodynamics ascribes infinite values to some physical quantities such as masses and charges, and these must be replaced by finite values before the calculation can be completed. The procedure for replacing these values, formulated by Feynman, Julian Schwinger, and others, is known as finite renormalization. As even admirers of quantum electrodynamics recognize, finite renormalization violates standard mathematical rules and has only an ad hoc justification.[11]

Largely because of its dependence on finite renormalization, quantum

10. Dirac (1980a), pp. 43–44. Similarly, A. M. Taylor (1966), p. 38, writes: "The elegant beauty of the theoretical edifice [of general relativity] is thought sufficient reason for believing it to be true." Further cases in which scientists have retained theories in virtue of their aesthetic properties, overruling apparent experimental results, are examined in Barrow (1988), pp. 345–352.

11. On the history of quantum electrodynamics, see Schweber (1994); on the development of finite renormalization, see Aramaki (1987).

electrodynamics has struck many physicists as aesthetically displeasing. Dirac was among those who refused on these grounds to accept it as close to the truth. He did not deny that quantum electrodynamics accounted well for empirical data; but he saw its inelegance as evidence that it had not captured the truth about phenomena. He wrote for example: "Recent work by Lamb, Schwinger, Feynman and others has been very successful [. . .] but the resulting theory is an ugly and incomplete one, and cannot be considered as a satisfactory solution of the problem of the electron."[12] On another occasion, in 1950, Dirac told Freeman J. Dyson, who had asked his opinion of recent developments in quantum electrodynamics: "I might have thought that the new ideas were correct if they had not been so ugly."[13]

Quantum electrodynamics was not the sole empirically successful theory that Dirac believed on aesthetic grounds to be distant from the truth. He rejected the nonlinear spinor theory of Heisenberg on aesthetic grounds: "My main objection to your work is that I do not think your basic (non-linear field) equation has sufficient mathematical beauty to be a fundamental equation of physics."[14] If aesthetic judgment were able to reveal that a theory is distant from the truth, scientists would no longer be deceived into endorsing incorrect theories by the false positives produced by empirical evaluations of theories.

The problem that empirical success does not demonstrate that a theory is true or close to the truth has been recognized for as long as science has been practiced. Medieval natural philosophers attempted to overcome it by means of a distinction between theses and hypotheses. On their definitions, a hypothesis is a set of propositions that saves the phenomena of a certain domain, but need not be true; a thesis is a theorem that is established scientifically in the Aristotelian sense of this term, or validly deduced from evident first principles and thus indubitably true. The acceptability of a claim as a hypothesis can be judged on empirical criteria, but theses must pass certain more stringent, nonempirical tests. In the scholastic version of this doctrine, these included a test of consistency with the works of Aristotle and with holy scripture. While the criteria envisaged by Dirac to reveal false positives in the empirical evaluation of theories are different from those of medieval natural philosophers, it is interesting to see two solutions of the same form being proposed at an interval of several centuries for the same problem.

12. Dirac (1951), p. 291.
13. Quoted from Kragh (1990), p. 184. Further evidence of Dirac's aesthetic dislike of quantum electrodynamics is provided in Shanmugadhasan (1987), p. 53, and especially in Kragh (1990), pp. 183–185.
14. Quoted from Brown and Rechenberg (1987), p. 148. The passage is contained in a letter of Dirac of March 1967.

To illustrate the conviction that aesthetic criteria can reveal the proximity of theories to the truth, I have so far used statements by Dirac. Another physicist who devoted great attention to the aesthetic properties of theories was Einstein. In fact, the dominant impression that Einstein left on many of his contemporaries was of his sensitivity to the beauty of theories. His son Hans Albert Einstein, also a physicist, said of him: "He had a character more like that of an artist than of a scientist as we usually think of them. For instance, the highest praise for a good theory or a good piece of work was not that it was correct nor that it was exact but that it was beautiful."[15] Moreover, Einstein seems to have thought that an aesthetic judgment of a theory can overrule an unfavorable empirical judgment of it, if Dirac is to be believed: "Einstein seemed to feel that beauty in the mathematical foundation was more important, in a very fundamental way, than getting agreement with observation."[16]

Einstein set out a systematic account of theory assessment in his autobiographical notes.[17] In his view, a scientific theory is a structure of interrelated concepts, which scientists formulate in the attempt to account for empirical data. A scientific theory therefore exhibits relations of two sorts: relations holding among the concepts that make up the theory, and relations between these concepts and bodies of data. A theory may be evaluated for both. There are thus two levels of theory assessment, as Einstein calls them: an internal level, on which judgment is passed on the theory's inherent conceptual structure, and an external level, on which judgment is passed on the relationship of the theory to empirical data.[18]

Einstein summarizes the requirements of the external level by the remark that "the theory must not contradict empirical facts."[19] He felt that the criterion of agreement with data has two shortcomings, however. Firstly, any discrepancies between a theory and data may be eliminated by means of ad hoc hypotheses, but these reduce the overall scientific worth of the theory.[20] Secondly, in modern physics the lines of reasoning by which predictions are drawn from the principles of a theory have be-

15. Quoted from Whitrow (1967), p. 19. Bondi agrees: "As soon as an equation seemed to him to be ugly, he really rather lost interest in it [. . .]. He was quite convinced that beauty was a guiding principle in the search for important results in theoretical physics" (ibid., p. 82). Further evidence of the importance that Einstein gave to the aesthetic properties of theories is given throughout Pais (1982).

16. Dirac (1982b), p. 83.

17. Einstein (1949).

18. For further discussion of Einstein's two levels of theory assessment see P. Barker (1981), pp. 138–142, and Miller (1981), pp. 123–131.

19. Einstein (1949), p. 21.

20. Ibid., pp. 21–23.

come long and convoluted, so that "the confrontation of the implications of theory by the facts becomes constantly more difficult and more drawn out."[21] For these reasons, physicists should not attach great weight to the outcomes of empirical tests of high-level theories—a precept with which Einstein notably complied.[22]

Theory assessment on the internal level "is not concerned with the relation to the material of observation."[23] Rather, judgments on this level pertain to the "naturalness" or logical simplicity of theories, as well as to the definiteness of their descriptions. In Einstein's view, the degree of a theory's logical simplicity is determined by the number of arbitrarily chosen axioms that it contains. To illustrate the idea, Einstein refers to Newtonian mechanics. In Newton's own formulation of it, Newtonian theory has the inverse square law of gravitation as an explicit axiom and the axioms of Euclidean geometry as implicit axioms.[24] The choice of an inverse square function of distance as the law of gravitation is arbitrary in Newton's formulation of the theory: it is not forced on us by the other axioms. However, as Einstein points out, there is an alternative formulation of the theory in which this choice is not arbitrary. The least-power spherically symmetric solution of Laplace's equation for a potential is an inverse first-power function of distance. Differentiation of this potential function yields a force law which is an inverse second-power function of distance. The choice of an inverse square function of distance as the law of gravitation is now forced upon us by the axioms of Euclidean geometry. In this alternative formulation, the theory has a greater degree of what Einstein calls logical simplicity.

In the account of theory assessment that he gives in his autobiographical notes, Einstein does not mention aesthetic criteria. However, there are good reasons for concluding that the criteria that Einstein envisaged being used on the internal level of theory assessment are aesthetic. First, the property that Einstein calls naturalness, for which theories are assessed on the internal level, is arguably an aesthetic property. Second, the supposition that the internal criteria are aesthetic would help to explain the importance that Einstein attached to beauty in theories. For example, under this supposition, Einstein's belief that assessments of theories on the internal level should be given great weight compared to

---

21. Ibid., p. 27.
22. Holton (1973), pp. 252–253, tells of Einstein's disregard for the empirical findings of Kaufmann, which seemingly refuted the special theory of relativity; Rosenthal-Schneider (1980), p. 523, tells of Einstein's confidence in the validity of the general theory irrespective of the outcome of Eddington's solar eclipse expedition.
23. Einstein (1949), p. 27.
24. Einstein presents this example ibid., pp. 29–33.

The Relation of Beauty to Truth

those on the external level would explain his attitude toward quantum theory, which—as we shall see in Chapter 11—he rejected on aesthetic grounds despite its empirical success.

4. The Properties of Theories and the Properties of Phenomena

The time has come to ask whether the conviction that aesthetic criteria have the power to reveal theories' proximity to the truth could ever be justified. In this section, we examine some attempts to justify it that do not succeed; in the next section we shall look at a more promising approach.

A defense of the claim that aesthetic criteria can reveal theories' proximity to the truth might take the following line. Aesthetic criteria evaluate theories for their possession of certain aesthetic properties. Phenomena too have aesthetic properties. Theories are likely to be close to the truth if they have the correct aesthetic properties, i.e., the same aesthetic properties as the phenomena that they attempt to describe. Therefore, as long as our aesthetic criteria for theory evaluation in a given domain express preference for the aesthetic properties that are shown by the phenomena of that domain, they can be expected to reveal which theories are close to the truth. For example, if phenomena of a certain domain have particular symmetry properties, our theories of them are more likely to be true if they show those same symmetries. Thus, as long as our aesthetic criteria are tuned to these symmetries, they amount to a test for proximity to the truth in this domain.[25]

Any persuasiveness that this argument appears to have is owed to a confusion between two sets of properties: the properties of scientific theories and the properties of phenomena. A property of an abstract entity such as a scientific theory and a property of a concrete entity such as a natural phenomenon cannot be the same property, regardless of their names. For instance, no symmetry property shown by a theory can be the same property as a symmetry property of a phenomenon. Because of this, the prescription to choose theories that have the aesthetic properties shown by their subject matter cannot be obeyed, even in principle. Indeed, in many cases this prescription is incomprehensible. For example, the gravitational field of a point mass has spherical symmetry, but what would it mean to demand spherical symmetry of a theory of gravitation?

Alternatively, we might consider defending the claim that aesthetic

25. The claim that theories are likely to be true if their symmetry properties replicate those of phenomena is put forward by, among others, Yang (1961), pp. 52–53.

criteria can reveal theories' proximity to the truth by arguing that a theory of given phenomena is likely to be close to the truth if it correctly describes (not replicates) the aesthetic properties of those phenomena. This line of argument is more persuasive. A true theory about the phenomena of a certain domain would be an accurate and complete account of them; therefore, if some phenomena had a particular aesthetic property, this fact would be reported by a true theory. However, this argument fails to specify how the aesthetic properties of theories can be used as an indicator of theories' proximity to the truth. A theory can correctly describe the aesthetic properties of phenomena without showing any aesthetic properties that are recognizably correlated with them. For instance, a true theory about snowflakes would doubtless report the fact that they have hexagonal symmetry, but need not show any particular aesthetic properties in order to do so. Therefore, ascertaining which theory most correctly describes the aesthetic properties of given phenomena is just an aspect of, and no easier than, ascertaining which is the empirically most adequate theory of those phenomena.

Of the physicists who believe that aesthetic criteria have the power to reveal theories' proximity to the truth, many seem to be persuaded of this by Maxwell's equations. As we saw in Chapter 3, Maxwell's equations show notable symmetries, in virtue of the fact that, for example, interchanging the electric and magnetic field intensities in the equations leaves their content nearly unchanged. Many physicists believe that a theory that lacked the symmetries of Maxwell's equations would be unable to account successfully for electromagnetic phenomena. Their argument appears to be that particular symmetries are shown by electromagnetic phenomena, and a theory that wishes to account for them must show the same symmetries. But this argument is invalid. All that a theory is obliged to do with respect to symmetry in order to account for empirical data on electromagnetic phenomena is to report correctly the symmetries that electromagnetic phenomena show. The particular symmetries that Maxwell's equations possess are an extra property of them, which they need not have in order to account for the data. In other words, to the extent that electromagnetic phenomena and an empirically adequate theory of electromagnetic phenomena can be described as showing similar symmetries, this accord should be regarded as fortuitous.

The defenses of the claim that aesthetic criteria can reveal theories' proximity to the truth that we have so far examined have a further defect: there is circularity in the idea of judging that a theory is close to the truth on the grounds that it correctly replicates or describes the aesthetic properties of phenomena. For in many cases our belief that the phenomena show particular aesthetic properties is based entirely on the theory whose proximity to the truth we are attempting to assess. It may be possible to discern by observation that snowflakes show hexagonal symmetry; but our only grounds for believing that electromagnetic phenomena show particular symmetries are that Maxwell's equations tell us that they do. It would therefore be illegitimate to cite the belief that electromagnetic phenomena have these symmetries in support of the claim that Maxwell's equations are close to the truth. In all such cases, the belief that the phenomena have particular aesthetic properties is an effect of choosing among given theories, not a basis for doing so.

I believe that appreciation of these points is embedded in scientists' practice, regardless of what they may claim about the accord of the aesthetic properties of theories with those of phenomena. A preference for theories that show certain forms of symmetry may entrench itself deeply in a community, but it is promptly disavowed if theories that show those symmetries cease to achieve empirical success, or if a theory with different symmetry properties demonstrates superior empirical virtues. In fact, the preference for a particular form of symmetry amounts merely to a community's retrospective weighting of the symmetry properties of theories that have previously demonstrated the greatest empirical success.

5. The Possible Success of the Aesthetic Induction

As we have seen, the claim that we can recognize the truth of a scientific theory by noting a correspondence between its aesthetic properties and those of phenomena cannot be sustained. However, it may still be that there are aesthetic criteria that are reliable indicators of the empirical adequacy of theories; and persistent use of the aesthetic induction is capable of revealing whether any such criteria exist and which they are.

Inductive inferences—that is, inferences from premises describing individual occurrences to a conclusion stating a universal regularity—lack the validity of deductive inferences, as David Hume pointed out in the eighteenth century.[26] Nonetheless, as twentieth-century writers such as Richard B. Braithwaite and D. H. Mellor have argued, there is a pragmatic justification for the policy of inductive projection.[27] This justification takes the following form. The world either shows or does not show regularities. If it does not show regularities, then no generalizations about contingent matters of fact are true, and there is also no policy that

26. Hume (1739).
27. Braithwaite (1953), pp. 255–292; Mellor (1988).

will deliver true generalizations about such matters of fact. In this case, to follow the policy of inductive projection is neither beneficial nor detrimental. If on the contrary the world does show regularities, then there are some true generalizations about contingent matters of fact. Inductive projection is at least as likely as any alternative procedure to discover these generalizations. This is because inductive projection is the policy of attempting to describe in the form of generalizations all patterns that occurrences appear to show, so that any genuine regularity of occurrences will certainly be captured in a generalization. Therefore, pursuing the policy of inductive projection is justified whatever degree of regularity the world shows.

This argument serves most obviously to justify the use of inductive projection in formulating scientific theories. But it justifies also the use of inductive projection in formulating precepts to guide action. If a precept has enduring effectiveness in promoting the achievement of a goal, it must be that there is a correlation between the actions recommended by the precept and the achievement of the goal. Thus, a precept that is effective owes its effectiveness to a regularity that the world shows. So the search for effective precepts, like the search for empirically adequate scientific theories, is a search for regularities in the world. We cannot be certain that the world shows the regularities that would be required for effective precepts to exist. If the world does not show such regularities, then no procedure for formulating precepts will deliver precepts that are effective. In this case, to follow the policy of inductive projection is neither beneficial nor detrimental. But if the world does show the regularities that are required for some precepts to be effective, then inductive projection will be at least as likely to discover these precepts as any alternative procedure. Therefore, formulating precepts by inductive projection is justified irrespective of the degree of regularity of the world.

More specifically, this argument vindicates the aesthetic induction as a procedure for formulating aesthetic criteria for theory choice. There may or may not be correlations between theories' having particular aesthetic properties and their having high degrees of empirical adequacy. If there are no such correlations, then no method of forming criteria for theory evaluation will identify any. But if some such correlations exist, then inductive projection will be at least as likely to discover them as any alternative procedure for formulating criteria.

This argument leaves open the question of whether aesthetic properties of theories that are correlated with high degrees of empirical adequacy actually exist. If there are no such aesthetic properties, then the aesthetic induction will continue to latch on to any aesthetic property that appears briefly to show a correlation with empirical adequacy, only to abandon it when the correlation fails to endure. In aesthetic space, the point representing the community's current theory will roam unrestrainedly rather than converge onto a particular location. In this case, there will never be found a justification for the conviction that certain aesthetic criteria have the power to reveal theories' proximity to the truth. If on the contrary there are aesthetic properties that are correlated with high degrees of empirical adequacy, then presumably a theory that has these properties will someday be formulated by a scientific community. Because such a theory will be bound to have empirical success, the aesthetic induction will ensure that the weighting of these properties in the aesthetic canon will increase, and therefore that the community's preferences will be biased toward theories that exhibit these properties. Any further such theories adopted on these criteria will also show empirical success, ensuring that the weightings of these properties continue to rise. In aesthetic space, the point representing the community's current theory will converge onto the location defined by this combination of aesthetic properties. In this case, the conviction that certain aesthetic criteria have the power to reveal theories' proximity to the truth will be found to be justified.

It is not yet clear which of these two states of affairs is realized in the world that we inhabit. Contrary to Dirac, Einstein, and others, I see little evidence that aesthetic properties correlated with high degrees of empirical adequacy in theories have yet been identified in any branch of science. If they had, the empirical benefit of choosing theories on particular aesthetic criteria would be far more obvious than it currently is. On the other hand, we have the assurance that the aesthetic induction is at least as likely as any other procedure to identify such aesthetic properties that may exist, and we cannot rule out that it will one day succeed in identifying some of them.

6. The Empirical Corroboration of Metaphysical World Views

In Chapter 3, I presented grounds for considering the allegiance that a scientific theory has to a metaphysical world view as one of its aesthetic properties. By considering metaphysical allegiance as an aesthetic property, I suggest that the model of the formation of aesthetic criteria that I have constructed applies equally to metaphysical criteria. In this section, I make explicit a few of the implications of this model for our understanding of metaphysical world views.

It is commonly believed that the claims made by metaphysical world views relate to states of affairs that lie outside the sphere of observable phenomena and therefore cannot be corroborated or refuted by empirical data. The etymology of the term "metaphysics" embodies this belief. If this belief were correct, which discoveries we make about the world and which scientific theories achieve the greatest empirical success would have no effect on our choice among alternative metaphysical world views. But it is not correct. As we saw in Chapter 3, metaphysical world views generate criteria stipulating which sorts of scientific theories are to be preferred. These criteria contribute to determine the choices among competing theories made by those who hold to a metaphysical world view. By evaluating the empirical performance of theories chosen in accord with some such criteria, we may judge how apt they are to recommend theories that are empirically successful. For instance, by recording the degrees of empirical success of theories satisfying the criteria set out by atomism, we can judge the criteria of atomism for their aptitude to recommend empirically successful theories. In this way, a metaphysical world view may acquire an empirical corroboration, as John Watkins has suggested.[28]

We may thus pose for metaphysical criteria a question analogous to the one that we discussed in the previous section for aesthetic criteria in general: is there a metaphysical allegiance that is more strongly correlated than any other with high degrees of empirical adequacy in scientific theories? If there were, there would exist a "scientific metaphysic," a metaphysical world view that has a uniquely strong empirical corroboration. My response to this question parallels the response that I gave to the question pertaining to aesthetic criteria in general. I doubt that a scientific metaphysic can be identified by, for example, noting a correspondence between the metaphysical properties of scientific theories and metaphysical properties of phenomena. But there is no obstacle to our identifying a scientific metaphysic by means of the aesthetic induction. This would involve giving weightings to metaphysical properties of theories in accord with the degrees of empirical success shown by theories that have these properties. It would become clear that a scientific metaphysic exists if scientists' choices among theories determined by the aesthetic induction converged onto theories that have a particular metaphysical allegiance.

Whether a scientific metaphysic exists has, I believe, been under investigation throughout the history of science. The investigation has taken place in, for example, the succession of metaphysical views that we examined in Chapter 3. Whether there is justification for the metaphysical view that attributes to inanimate objects such properties as active pow-

28. Watkins (1958).

The Relation of Beauty to Truth

ers, occult qualities, and a capacity for action at a distance was investigated by the scientific community in its progress from astrological, alchemical, and magical theories to Cartesian corpuscular mechanics and eventually to Newtonian gravitational theory.

I believe that, just as it is not yet clear that there is a combination of aesthetic properties in general that is correlated with high degrees of empirical adequacy in theories, so it is not yet clear that there is a metaphysical allegiance of theories that shows such a correlation. Nevertheless, the aesthetic induction is at least as successful as any other procedure at identifying a scientific metaphysic, and we cannot rule out that it will someday succeed in identifying one.

A Study of Simplicity

In scientific thought we adopt the simplest theory which will explain all the facts under consideration and enable us to predict new facts of the same kind. The catch in this criterion lies in the word "simplest." It is really an aesthetic canon such as we find implicit in our criticisms of poetry or painting. The layman finds such a law as $\partial x/\partial t = \kappa(\partial^2 x/\partial y^2)$ less simple than "it oozes," of which it is the mathematical statement. The physicist reverses this judgement.

        —J. B. S. Haldane, "Science and Theology as Art-Forms"

1. THE CONTROVERSY ABOUT SCIENTISTS' SIMPLICITY JUDGMENTS

Most philosophical accounts of theory choice acknowledge that, presented with two theories that are on other grounds equally worthy, scientists will prefer the theory whose claims are, in some sense, simpler. It is well known for example that physicists prefer to adopt simple statements rather than complicated ones as laws of nature. However, there is no agreement among philosophers about the nature of the simplicity considerations to which scientists appeal. One group of writers believes that the simplicity of a theory's claims is diagnostic of the theory's future empirical success and that simplicity considerations should therefore be viewed as an empirical criterion in theory choice. A second group claims that the simplicity of a theory is not correlated with its empirical performance—this is self-evidently true, some argue, since simplicity is an observer-relative property that different observers will find present in theories to different degrees. Some members of this second group maintain further that the simplicity of a theory is an aesthetic property of it, and that scientists' recourse to simplicity considerations therefore amounts to an appeal to aesthetic criteria. In this chapter, I try to resolve the controversy by advancing a new interpretation of the simplicity considerations to which scientists appeal in theory choice.[1]

Those who wish to portray scientists' simplicity considerations as an empirical criterion for theory choice must establish that the simplicity properties of theories have some correlation with their empirical success.[2] The claim of this sort that is most often defended is that, of two theories that fit available empirical data equally well, the simpler theory is empirically superior. There are three popular arguments for this claim: I call these the argument from the simplicity of the phenomena, the argument from informativeness, and the argument from likelihood.

The argument from the simplicity of the phenomena takes the following form: since the phenomena are simple, a theory about a given phenomenon is more likely to be empirically adequate if it too is simple. This argument has two principal defects. First, since judgments of simplicity are relative rather than absolute, the claim that the phenomena are simple, like the related claim that nature is uniform, is not well formulated. One would have to claim rather that the phenomena are simple compared to some other entity, but it is difficult to see what entity could act as a worthwhile term in this comparison. Second, our only grounds for believing that a given phenomenon is to some degree simple are our theories about that phenomenon. Therefore, it is illegitimate to cite the belief that the phenomena are simple in support of the claim that a given theory is empirically adequate.[3]

The argument from informativeness has two premises. The first premise says that the simpler of a pair of theories is the more informative; the second premise says that the more informative of two theories is the empirically superior. From this it follows straightforwardly that, of two theories, the simpler is empirically superior. A typical defense of the first premise is given by Stephen F. Barker: "If one system is simpler than another [. . .] then the simpler one 'says more,' it has 'more content,' because it excludes a greater number of possible models; therefore it runs more risk of being contradicted by the evidence. A system which takes a risk yet survives deserves more credit, it earns more credibility, than does a system which survives but says less and thus has taken less risk."[4] The second premise is supported for example by Elliott Sober: "The

1. This chapter is developed from McAllister (1991b).
2. Salmon (1961), pp. 275–276, discusses the kinds of link that might hold between the simplicity of a theory and its truthlikeness or empirical adequacy.
3. The relation between claims that the phenomena are simple and methodological principles of simplicity is discussed in Sober (1988), pp. 37–69.
4. S. F. Barker (1957), pp. 181–182. Barker's explanation of the value of simplicity in theory preference would win the agreement of Popper: see Popper (1959), pp. 140–142.

more informative our knowledge claims are about the properties of the individuals in our environment, the less we need to find out about the special details of an arbitrary individual before we can say what its properties are."[5] On the argument from informativeness, therefore, the simpler theory is a superior predictive tool by dint of being, on its own, more informative than a more complex theory would be.

The argument from likelihood rests on the claim that, of two theories that fit the empirical data equally well, the simpler has a higher likelihood of being true. This claim is generally defended by appeals to the Bayesian account of theory confirmation: this account, named after the eighteenth-century probability theorist Thomas Bayes, suggests that simpler theories yielding the same predictions as more complex ones derive stronger support from any common body of favorable evidence.[6] On this view, simpler theories are empirically superior than more complex theories in the sense of being worthier of belief. The claim that a simpler theory is better supported than a more complex one by any common favorable evidence is occasionally advanced in scientific practice: George C. Williams, for instance, argued in evolutionary biology that the theory of genic selection and organic adaptation is better supported by the data than the theory of group selection and biotic adaptation, because it is simpler.[7]

If any of these arguments were sound, it would be legitimate to use degree of simplicity as an empirical criterion for theory choice alongside, for example, the criterion of consistency with extant empirical data.

In opposition to these arguments there stands the view that the simplicity of a theory is not correlated with its empirical performance. This view is endorsed by Newton-Smith, who writes: "There is no reason to see greater relative simplicity [. . .] as an indicator of greater verisimilitude" in a theory.[8] Gerd Buchdahl shares this opinion, listing "maxims of simplicity and economy" alongside "general metaphysical notions" among the extraempirical criteria that scientists use.[9] This view is defended usually by the claim that simplicity is an observer-relative property, so appraisals of the simplicity of theories cannot measure an objective quality of theories such as their degree of empirical success or

5. Sober (1975), p. 3.
6. For the Bayesian account of simplicity, see Kemeny (1953) and Rosenkrantz (1976).
7. Williams (1966), pp. 123–124.
8. Newton-Smith (1981), p. 231. The existence of any correlation between the simplicity of a theory and its verisimilitude has been denied also by Bunge (1963), pp. 96–98, and Harré (1972), p. 45.
9. Buchdahl (1970), p. 206.

A Study of Simplicity

of proximity to the truth. There are two common arguments for this claim.

The first argument exploits the indeterminacy of the expression "the simpler of two theories." Suppose that we require a theory to account for a certain body of data; that several theories are available, each of which accounts for the data by means of a different polynomial function, or function of the form $y = a + bx + cx^2 + \ldots$; and that these theories accord equally well with the data and are equally worthy in all other respects. We decide to choose the theory that puts forward the simplest polynomial. As Rom Harré points out, there are many different simplicity criteria that we may apply in this case.[10] They include the following:

1. The criterion of the number of variables, which stipulates that the simplicity of a polynomial varies inversely with the number of its independent variables—so that a polynomial in x alone is simpler than one in x and z;

2. The criterion of the magnitude of exponents, according to which the simplicity of a polynomial varies inversely with the magnitude of the highest exponent that appears in it—so that a polynomial in which the highest-exponent term is x^2 is simpler than one containing the term x^3;

3. The criterion of integer exponents, which stipulates that a polynomial containing only integer exponents is simpler than one in which some exponents are nonintegers. On this criterion, Newton's law of gravitation, $F = Gm_1m_2/r^2$, is simpler than the possible alternatives in which the exponent of the distance differs slightly from 2, such as $F = Gm_1m_2/r^{2.01}$. The latter claim has been advanced by many physicists, most notably Laplace.[11]

All these simplicity criteria are of equal intrinsic worth, since it is intrinsically no more meritorious for a polynomial to be simple in one of these respects than to be so in another. This means that it is impossible to determine on a nonarbitrary basis which is the simplest of, say, $ax + bz$, ax^6, and $ax^{2/3}$. The injunction to choose the simplest polynomial therefore fails to resolve this case of theory choice. In the more general case of theory choice, in which the competing theories differ in more respects than the form of a polynomial, the number of alternative kinds of simplicity increases greatly. Since all these kinds of simplicity are of equal intrinsic worth, any judgment that one theory is simpler than another is arbitrary. Simplicity considerations—this argument concludes—are thus

10. Harré (1960), pp. 138–139.
11. Laplace (1813), 2:10–11.

not suited to picking from among a number of competing theories the one that is closest to the truth.[12]

The second attempt to establish that simplicity is an observer-relative property hinges on what is actually meant by the statement that a theory is simple. Harré claims that frequently this statement expresses no more than that the theory is familiar to the speaker: "In many cases when a theory is judged to be simple attention is not being drawn to the paucity of concepts employed in its construction or to the simplicity of its structure but to the fact that the model which it is based upon is one [with] which either the author of the theory or preferably everyone, is quite familiar."[13] For instance, classical physicists might regard the kinetic theory of gases as simple because of their familiarity with Newtonian mechanics. The degree of familiarity that a community has with a particular model is clearly observer relative. If evaluations of how simple a theory is are indeed determined by the familiarity of the model on which it is based, then simplicity considerations cannot be relied upon in theory choice to select the theory that is closest to the truth.

Lastly, according to some writers, a theory's simplicity properties are among its aesthetic properties.[14] For instance, Einstein appears to have believed that, in the words of Yehuda Elkana, "simplicity was equivalent to beauty" in theory choice.[15] This view gains plausibility once we observe that simplicity properties are capable of giving rise to the sense of aptness that constitutes our criterion for recognizing a theory's aesthetic properties.

2. Simplicity and the Unification of Phenomena

The notion of simplicity is closely related to that of unifying power: scientific theories may be called simple to the extent that they establish the unity of phenomena that were previously considered distinct. The power

12. An argument to the effect that there are many conceivable kinds of simplicity, and none to which one can appeal on objective grounds in theory evaluation, is briefly pursued also by Priest (1976), pp. 436–437.

13. Harré (1960), p. 143; the suggestion that the most familiar construction will be the one judged the simplest is made also by Priest (1976), p. 437.

14. Among philosophers who regard scientists' simplicity criteria as aesthetic are Harré (1960), p. 147, Walsh (1979), and several of the contributors to Rescher (1990).

15. Elkana (1982), p. 222. Among scientists, Tsilikis (1959) and E. O. Wilson (1978), p. 11, also regard the simplicity of theories as an aspect of their beauty. Lamouche (1955), pp. 81–132, and Derkse (1993) offer further examples of scientists' treatment of simplicity considerations as aesthetic.

of theories to unify domains of phenomena is often regarded as a purely empirical capability. On this basis, some might expect the link with unifying power to allow us to formulate a notion of simplicity that avoids any mention of aesthetics.[16] For example, Maxwell's unification of optics and electromagnetism is frequently portrayed as a purely empirical achievement: on this premise, one might conclude that no aesthetic judgment is involved in recognizing the simplicity properties of Maxwell's theory. Alas, this belief cannot be sustained: the notion of unifying power shows the same indeterminacy and aesthetic aspects shown by the notion of simplicity. There are many ways in which classes of phenomena can be said to admit unification. Because of this, the prescription that scientists should choose the theory with the greatest unifying power is indeterminate, like the prescription that they should choose the simplest of several polynomials. Moreover, each of these different forms of unification may be attributed aesthetic value. So if a scientist has particular aesthetic preferences, his or her choice between theories that perform different kinds of unification will be determined partly by aesthetic considerations.

The claim that unification of phenomena may be performed in different ways is substantiated by the history of physics. Newton's program for unifying physics involved analyzing all physical phenomena as the manifestations of central forces whose magnitude varies with distance. Each of these forces was to be described by a causal law, modeled on the law of gravitation. The repeated and combined application of these laws was expected to solve all physical problems, unifying celestial mechanics with terrestrial dynamics and the sciences of solids and of fluids. Simultaneously, however, Leibniz was proposing to unify physical science in a different way: on the basis of abstract and fundamental principles governing all phenomena, such as the principle of continuity, the principle of conservation of force, and the principle of relativity of motion. In the Newtonian program, the unity of the physical world derives from the fact that causal laws of the same form apply to every event in it; in Leibniz's program, it derives from the fact that a few universal principles apply to the universe as a whole.

As Newton and Leibniz themselves perceived, these forms of unification are alternatives: the considerations that establish the unity of the physical world in the Newtonian program are irrelevant to the Leibnizian program, and vice versa. So although both a Newtonian and a Leibnizian theory may be said to have great unifying power and therefore

16. The simplicity of theories is linked to their unifying and explanatory power by, for example, Friedman (1974) and Kitcher (1989), pp. 430–447.

great simplicity, one cannot univocally be declared superior in this respect to the other. This is shown too by the subsequent development of the two programs. In the eighteenth and nineteenth centuries, the Newtonian program was the more widely pursued: Roger Boscovich and the French school of rational mechanics had much success in unifying phenomena by means of casual laws. Herman Boerhaave, the champion of Newtonianism at Leiden, saw the brand of simplicity embodied in Newtonian theory as a sign that it had uncovered the truth about phenomena: it is Newtonian simplicity to which he refers in his motto, *Simplex veri sigillum* ("Simplicity is the emblem of truth").[17] But more recently the Newtonian approach has become less fashionable: twentieth-century physicists have striven to unify physical phenomena not primarily by accumulating causal laws but by formulating great conservation and symmetry principles reminiscent of those of Leibniz. There is no objective basis for saying that the theories produced by either approach have had greater unifying power.

Furthermore, each of these approaches has distinct aesthetic properties, which may lead physicists to prefer one to the other. On the one hand, as Hutcheson noted, it is not difficult to discern aesthetic value in Newton's law of gravitation. Many physicists have expressed an aesthetic preference for theories that unify phenomena by citing causal laws—Helmholtz is an example.[18] On the other hand, aesthetic considerations played an important role in Leibniz's principle thinking, and many present-day physicists find it pleasing to regard physical science as based on a few conservation and symmetry principles.[19] So appeal to the notion of unifying power establishes neither that the simplicity of theories is a purely empirical property of them nor that the notion of simplicity has no aesthetic aspects.

3. DEGREES AND FORMS OF SIMPLICITY

There are thus three views of scientists' simplicity criteria in circulation: as diagnostic of theories' empirical adequacy, as observer-relative criteria for theory assessment, and as aesthetic criteria. Up to now, most philosophers have regarded the first option as excluding the latter two, and vice versa. For instance, Sober sees his model of simplicity criteria, which por-

17. Lindeboom (1968), pp. 268–270 and plate 25.
18. On Helmholtz's use of aesthetic criteria in theory appraisal, see Hatfield (1993), pp. 553–558.
19. The role of aesthetic considerations in Leibniz's natural philosophy is documented by Boullart (1983) and Breger (1989, 1994).

A Study of Simplicity

trays them as diagnostic of theories' empirical adequacy, as one that "explains away the faulty intuition that aesthetic simplicity is involved in hypothesis choice."[20] The belief that these options are mutually exclusive is founded presumably on an assumption that there is only one simplicity criterion that scientists use. As one might argue, if the degree of a theory's simplicity is diagnostic of its future empirical success, one would not want to consider judgments of its simplicity as anything other than empirical evaluations; and if the appraisal of a theory's simplicity is observer relative, one would not want to consider its outcome as revealing the theory's empirical worth. Hans Reichenbach is almost alone among philosophers of science in suggesting that scientists use both empirical and aesthetic criteria of simplicity, and he envisages the aesthetic criterion being used only to determine which of two logically equivalent theories has the more convenient form.[21]

The belief of philosophers that these options are incompatible feeds into their reconstructions of scientists' acts of theory evaluation. For instance, Donald J. Hillman portrays the scientific community as divided into two camps, one regarding simplicity as nothing but an empirical criterion and the other applying it in purely aesthetic judgments: "It is likely that [some scientists] will choose the simpler theory as better supported by the evidence, even though both theories are equally compatible with the evidence in their favor. [. . .] Other practitioners, however, feel that the notion of simplicity cannot be helpfully characterized. Simplicity is, in their opinion, much too heavily dependent on aesthetic and pragmatic considerations to be genuinely analyzable."[22]

Contrary to this belief, I now show that the view of simplicity considerations as diagnostic of a theory's empirical adequacy and the view of them as aesthetic can be combined into a richer notion of what it is for a theory to be simple.

Suppose that a theory's simplicity can be described completely by specifying the values of some parameters. How many parameters must be fixed in order to provide an exhaustive description of the simplicity of a given theory? It is not enough to specify a degree of simplicity for the theory. Consider the following facts. A physical theory may show numerical simplicity, as Dirac wished, in virtue of attributing simple values to coefficients and exponents. It may show explanatory simplicity, as

20. Sober (1975), p. 172. For a similar claim, see Sober (1984), p. 238 n. 16. Feuer (1957), pp. 115–117, also denies that the appraisal of a theory's simplicity can possess aesthetic aspects as well as holding clues to the likely future empirical success of the theory.
21. Reichenbach (1938), pp. 373–375.
22. Hillman (1962), pp. 225–226.

Newtonian physicists wished, in virtue of adducing the same explanatory laws for a wide range of phenomena. It may show ontological parsimony, as Ernst Mach wished, in virtue of postulating a small number of different material entities.[23] It may show logical simplicity, as Einstein wished, in virtue of resting upon a small number of independent postulates.[24]

Let us call each of these respects in which theories may be simple a form of simplicity. There are very many forms of simplicity that theories may exhibit. It is even possible to distinguish subforms within the four forms that I have just listed. For example, there are distinct forms of numerical simplicity, as is shown by our earlier examination of the ambiguity of the expression "the simplest polynomial." Similarly, David Lewis distinguishes between two kinds of ontological parsimony: a theory is, he writes, qualitatively parsimonious if it posits a small number of fundamentally different sorts of entities, and quantitatively parsimonious if it minimizes the number of instances of the entities of the sorts that it posits.[25] On my account, each of these is a form of simplicity.

The degree to which a theory displays one form of simplicity is uncorrelated with the degree to which it shows another. A theory may achieve one degree of simplicity by having equations that contain only simple numbers, achieve a second degree by reducing a given range of phenomena to a common explanatory schema, attain some other degree of simplicity by postulating a small number of material entities, and exhibit yet another degree by resting on few postulates. We should certainly say that each of these attainments is a component of the theory's simplicity. Thus, a full description of the simplicity of a theory must specify the degree to which it exhibits each of the forms of simplicity that may be envisaged in theories.

So far, we have been considering the problem of assembling a full description of the simplicity that a theory in fact exhibits. Let us now

23. Mach's preferences concerning the simplicity of theories arise from his overall view of science: "Science [. . .] may be regarded as a minimal problem, consisting of the completest possible presentment of facts with the *least possible expenditure of thought*" (1883, p. 586). Mach's criterion of simplicity is further discussed in Ray (1987), pp. 1–50.

24. Commenting upon a discrepancy of up to 10 percent between the measured value of a gravitational deviation of a light ray and the magnitude of the effect calculated from general relativity, Einstein weighed structural simplicity against any empirical deficiency of the theory: "For the expert, this thing is not particularly important, because the main significance of the theory does not lie in the verification of little effects, but rather in the great simplification of the theoretical basis of physics as a whole" (quoted from Holton 1973, p. 254). Further discussion of Einstein's appeal to simplicity criteria is offered in Hesse (1974), pp. 239–255, and Elkana (1982); on this topic see also Williamson (1977).

25. Lewis (1973), p. 87.

A Study of Simplicity

proceed to the problem of determining the number of parameters that we must fix in order to specify the simplicity that we wish to see exhibited by theories. This problem is of course the one that we face when we wish to state a preference among theories on grounds of simplicity.

A specification of the simplicity that we wish to see exhibited by theories need not be as extensive as a complete description of the simplicity actually exhibited by a given theory. While in the latter case we must specify the degree to which the theory exhibits each of the forms of simplicity that theories may possess, in the former case we need specify only the degree to which theories ought to exhibit each of a limited number of forms of simplicity, viz., those forms to which we accord particular value. In fact, a scientist who expresses a preference among theories in respect of their simplicity properties will typically mention only one form of simplicity: for example, Mach attached value only to ontological parsimony.

Even so, the shortest possible specification of the simplicity that a scientist wishes theories to show must fix the values of two parameters: the form of simplicity that he or she wishes to see in theories, and the degree to which theories should show that form of simplicity. As before, these two parameters are independent of one another: expressing the wish for theories to show ontological parsimony does not entail a wish to see theories show this form of simplicity to any particular degree. In most cases of theory choice, scientists are engaged in ascertaining not how well an isolated theory satisfies their criteria but which among several theories does so best. Still, even the latter task requires both a criterion of form and a criterion of degree of simplicity. A scientist who must choose on grounds of simplicity between theories exhibiting diverse forms of simplicity to varying degrees will need to identify both a privileged form of simplicity and a degree to which theories should show that form.

One criterion of degree of simplicity is more popular than any other. Most scientists in most circumstances prefer a theory showing a greater degree of simplicity to one showing a lesser degree. But the contrary preference is not unknown. For example, among scientists who have paid regard to ontological parsimony in assessing theories, some have preferred theories to show this form of simplicity to lesser degrees. Nowadays this preference is expressed most frequently in particle physics. In cases where fundamental principles and empirical data are compatible both with the existence and with the nonexistence of a particular hypothetical particle, some physicists prefer theories to affirm its existence. Dirac once expressed preference for a theory that asserted the existence of the magnetic monopole, saying that, as long as it was consistent with deep physical principles and available data, "one would be surprised if Nature had made no use of it."[26] Similarly, some physicists have preferred theories to affirm the existence of tachyons, or particles capable of traveling faster than light.[27] This criterion of minimal simplicity perhaps descends from the principle of ontological plenitude, an enduring metaphysical tenet that arose with Plato and was developed by Leibniz, according to which the range of actual entities exhausts the space of potential being.[28] The fact that there can be a preference for less simple theories as well as a preference for simpler ones demonstrates that a criterion of degree of simplicity underlies, explicitly or implicitly, all appeals to simplicity considerations in theory choice.

Whatever criterion of degree of simplicity a scientist adopts, it does not in general yield a univocal recommendation in a case of theory choice unless the scientist also adopts a criterion of form of simplicity. Suppose that a group of scientists, all of whom agree that greater degrees of simplicity are preferable to lesser degrees, is faced with several competing theories. If the group has no preference among alternative forms of simplicity, their common preference for greater degrees of simplicity will fail to compel any choice among the theories. In general, it will be possible to portray each of the available theories as the simplest, on the grounds that there is some form of simplicity that it exhibits to a greater degree than its competitors. As Imre Lakatos put it, "Simplicity can always be defined for *any* pair of theories T_1 and T_2 in such a way that the simplicity of T_1 is greater than that of T_2."[29]

This analysis sheds light on, among other things, Ockham's razor. Some scientists and philosophers seem to believe that applying Ockham's razor in theorizing and theory assessment is self-evidently justified: in fact, Ockham's razor is merely the statement of a particular preference among forms of simplicity. Its classic formulation, *Frustra fit per plura, quod potest fieri per pauciora* ("It is vain to do with more what can be done with less"), admits two interpretations. One recommends ontological parsimony, and amounts to *Entia non sunt multiplicanda praeter necessitatem* ("Entities are not to be multiplied beyond necessity"); the other recommends moderation in postulating explanatory principles. Therefore, if it is to be determinate, Ockham's razor must firstly be accompanied by a specification of which of these forms of simplicity—

26. Quoted from Kragh (1990), p. 214.

27. On the tendency to postulate the existence of tachyons, see Kragh (1990), p. 272. Kragh offers further examples of the use of the criterion of minimal simplicity in theory assessment ibid., pp. 270–274.

28. The history of the principle of plenitude is retraced by Lovejoy (1936).

29. Lakatos (1971), p. 131 n. 106.

A Study of Simplicity

ontological parsimony or explanatory economy—should be maximized. But even then, whichever form of simplicity is thus specified is merely one form among many that exist and to which value might be attached. Unless striving for ontological parsimony or explanatory economy is shown to promote better theories than other forms of simplicity, Ockham's razor has no claim to special status among simplicity principles.

The need to select a form of simplicity before theories can be ranked for degree of simplicity arises in many cases of theory choice. Imagine proposing to choose on grounds of simplicity between Nicholas Copernicus's theory of the solar system, which described planetary orbits as combinations of circles, and that of Johannes Kepler, which portrayed them as ellipses. Copernicus's theory may be considered the simpler in view of the fact that specifying a particular ellipse requires two parameters (the lengths of the axes) while specifying a particular circle requires only one (the length of the radius). On the other hand, Kepler's theory may be considered the simpler in virtue of the fact that the number of ellipses that it needs in order to account to given accuracy for the trajectory of a planet is smaller than the number of circles required by Copernicus's theory. In this case, as in general, the precept to choose the simpler theory of the two yields a determinate outcome only under a stipulation of the form of simplicity to which preference is to be attached.

There are many documented episodes in which scientists have chosen between theories on simplicity grounds. In every such episode there was either tacit agreement among the participants about the form of simplicity to which regard should be paid, or an explicit discussion of which form of simplicity was most fundamental. As Gerald Holton recounts, "Einstein and Planck debated strongly in 1914 whether the simplest physics is one that regards as basic *accelerated* motion (as Einstein had come to believe) or *unaccelerated* motion (as Planck insisted)."[30] As long as this difference of opinion persisted, even if both protagonists had obeyed the precept of choosing the theory that they regarded as the simpler, they would have failed to adopt the same theory. The principle of equivalence of inertial reference frames eventually persuaded physicists that the theory that regards accelerated motion as basic is simpler than the alternative.

Similarly, without a preference among forms of simplicity, one cannot propose choosing on simplicity criteria between the theories of gravitation of Newton and Einstein. The dilemma is stated by Dirac: "One of the fundamental laws of motion is the law of gravitation which, according to Newton, is represented by a very simple equation, but, according to Ein-

---

30. Holton (1978), p. 299 n. 8.

stein, needs the development of an elaborate technique before its equation can even be written down. [. . .] From the standpoint of higher mathematics, one can give reasons in favour of the view that Einstein's law of gravitation is actually simpler than Newton's."[31] A further respect in which Einstein's theory of gravitation might be judged conceptually simpler than Newton's is in its treatment of the notion of mass. Newtonian theory defines two quantities bearing the name "mass": gravitational mass, which appears in the law $F = Gm_1m_2/r^2$, and inertial mass, which appears in the law $F = ma$. Experiments suggest that the gravitational mass of a body is equal to its inertial mass. Newtonian theory has no resources to explain this equality and therefore portrays it as fortuitous; in Einstein's theory, gravitational and inertial mass are identical for deep-seated considerations. Because of this feature, physicists whose simplicity criterion prescribes that theories should leave unexplained as few coincidences as possible would find Einstein's theory simpler than Newton's.

The dependence of simplicity judgments on a criterion of form of simplicity also becomes apparent when we attempt to adjudicate on simplicity grounds between philosophical theories. Consider the dispute between scientific realism, which interprets the theoretical terms used by well-corroborated scientific theories as referring to actual entities, and instrumentalism, which regards them as notions useful in calculations but as having no referent in the world. One might argue that instrumentalism has greater ontological parsimony in virtue of the fact that it requires belief in fewer entities. On the other hand, scientific realism can claim to have greater explanatory economy, since it allows disparate observable phenomena to be explained as effects of a smaller number of hidden causes. We will be unable to choose between scientific realism and instrumentalism on simplicity grounds unless we have a preference between ontological parsimony and explanatory economy.[32]

The need for criteria of both form and degree of simplicity applies in all cases in which one seeks to rank objects according to simplicity. For instance, the skulls of higher vertebrates are made up of fewer separate bones than are the skulls of lower vertebrates, and may be regarded on this basis as simpler. On the other hand, in higher vertebrates these bones show structures such as fossa, crests, and processes, while in lower vertebrates they are featureless: in this respect, the skulls of higher vertebrates

31. Dirac (1939), p. 123. As Dirac observes, holding the view that Einstein's theory is simpler than Newton's "involves assigning a rather subtle meaning to simplicity."

32. The indeterminacy of simplicity criteria in the realism-instrumentalism debate has been noted by Rescher (1987), pp. 53–54.

might be regarded as more complex. An evolutionary biologist who wished to rank vertebrates according to the simplicity of their skulls would first have to specify which of these forms of simplicity should provide the basis for the ranking.[33]

Lastly, it may be noted that simplicity is not the only property for which a form must be specified before judgments of degree can be passed: another is similarity. In this parallel, allowance must be made for the fact that "is similar to" is a two-place predicate, while "is simple" is a one-place predicate. Is a tiger more similar to a zebra than a zebra is to a horse? The answer depends on the respect under which similarity is assessed. If we assess similarity in the matter of being stripy, the proposition is true; if we assess it in respect of the possession of an equine morphology, it is false. The respect under which similarity is assessed constitutes a criterion of, as one might say, form of similarity; once and only once a particular form of similarity has been stipulated can degrees of similarity be assessed. Considerations of this kind contributed to the decline of phenetics, the school in biological systematics that proposed to group organisms into taxa in the light of their overall morphological similarity. Its successor, cladistics, compares organisms not for their overall similarity but for their possession of shared individual character states: determining the character states on which two given organisms should be compared amounts to stipulating a criterion of form of similarity for that comparison.

4. Quantitative Definitions of Simplicity in Theory Choice

In recent years, several quantitative algorithms for the evaluation of scientific theories have been proposed. According to their advocates, these algorithms compute the degree of support that a body of empirical data affords to competing theories and thus identify which theory has the highest degree of empirical adequacy. If these claims were correct, such algorithms would yield objective evaluations of theories on empirical grounds.

In these algorithms, simplicity considerations have a prominent place: many of the algorithms incorporate the precept that, from among all theories on offer, the one that should be preferred is the simplest that meets certain requirements. They must therefore specify how the simplicity of

33. For further discussion of simplicity rankings of biological entities, see Levins and Lewontin (1985), pp. 16–18.

a theory may be quantified. Let us consider how this is done in two algorithms that have been proposed.

According to certain results in information theory, owed chiefly to Andrej N. Kolmogorov, the complexity of any message can be expressed as the length of the shortest possible description of it, or its minimal description, in a particular language.[34] For instance, the complexity of a string of numerals such as a telephone number is the length of the shortest specification that allows the string to be reconstructed. The minimal description of a string of random numbers is identical to the string itself, but the minimal description of a string that shows a pattern may be much shorter. Messages can then be ranked according to the length of their minimal descriptions in a particular language. Let us call such a ranking the Kolmogorov ranking of the given messages.

These ideas can be applied to scientific theories. We could associate to each theory a minimal description, the shortest specification required to generate the statements of the theory. The length of this description would be a measure of the theory's complexity. Save for its dependence on the choice of language, the length of the minimal description of a theory would be objective: it would not depend on the judgment of observers. The Kolmogorov ranking of a set of theories could then be constructed. Once this had been done, the injunction to choose the simplest of all available theories satisfying certain requirements would have objective content: no subjective decisions would be involved, and all observers who understood the principles underlying the ranking would find the same theory preferable.

Another quantitative algorithm for theory evaluation has been proposed by Paul Thagard. This consists of a computer program which constructs a quantitative measure of the worth of theories from various parameters. One of these is a measure of the simplicity of a theory T, which is defined, except in two cases, as follows:[35]

$$\text{Simplicity of } T = 1 - (\text{number of cohypotheses of } T)/(\text{number of facts explained by } T)$$

The two exceptions occur if a theory's cohypotheses are more numerous than the facts that it explains (in which case its simplicity is set at zero) and if a theory explains no facts (its simplicity is undetermined). Thagard defines cohypotheses of T as the auxiliary hypotheses that must be con-

---

34. A survey of Kolmogorov complexity theory and its applications in the evaluation of scientific theories will be found in Li and Vitányi (1992).
35. Thagard (1988), p. 90.

A Study of Simplicity

joined to T in order for its explanations to be accomplished. The measures of simplicity calculated from this formula, like those delivered by Kolmogorov's definition, can be used to rank any given theories by simplicity. This ranking too would be objective, since its construction would not rely on anything like the aesthetic taste of scientists.

What are the implications of the existence of quantitative definitions of simplicity such as those of Kolmogorov and Thagard? It might be thought that such definitions demonstrate that it is possible to choose among theories on simplicity grounds without attaching privilege to any particular form of simplicity, contrary to my conclusion in the previous section. After all, each of these definitions seems to yield a unique and objective ranking of theories, in which any given theory finds its place regardless of considerations about alternative forms. But this is not so: the existence of definitions of simplicity such as those of Kolmogorov and Thagard does not remove the need to attach privilege to a particular form of simplicity. Once a scientist has resolved to define the degree of simplicity of a theory as Kolmogorov suggests, he or she indeed has no need of a criterion of form of simplicity to perceive one theory as simpler than another. But there are many quantitative definitions of the simplicity of theories that the scientist might have preferred to Kolmogorov's. For instance, the scientist might have defined degree of simplicity as the number of discrete substances postulated by a theory or the number of its axioms. Any of these definitions would have yielded a ranking by simplicity. On what criterion did the scientist pick Kolmogorov's definition of simplicity rather than one of these alternatives? Each of these definitions of degree of simplicity corresponds to a form of simplicity. Therefore, the choice among such definitions is made on a criterion of form of simplicity.

This shows that the availability of quantitative definitions of the simplicity of theories does not make it less necessary for scientists to express preferences among forms of simplicity as well as among degrees of it. The quantitative approach taken by Thagard and others in philosophy of science, despite its seeming objectivity, does not show that there is no role to be played by aesthetic judgment in the evaluation of theories on simplicity grounds.

5. SIMPLICITY, BEAUTY, AND TRUTH

As the analysis of section 3 of this chapter suggested, concealed within generic talk of the simplicity of theories are in fact two simplicity criteria: degree and form. The problem, discussed frequently in the literature, of determining whether "the simplicity criterion" of scientists is a criterion diagnostic of a theory's empirical adequacy, an observer-relative criterion, or an aesthetic criterion is thus incorrectly formulated. The problem, we now see, is rather one of determining separately whether each of the two criteria is empirical, observer relative, or aesthetic. Beyond that, we will face the task of elucidating what implications our findings have for simplicity considerations that, like those that scientists actually make, appeal to both criteria jointly.

Let us first discuss the possibility that the criteria of degree and of form of simplicity are aesthetic. In disagreement with Sober and others, I think it is beyond doubt that, when scientists regard theories, some aesthetic pleasure is afforded to them by the theories' simplicity properties. The question is, how is the cause of this aesthetic pleasure distributed over scientists' perception of forms and of degrees of simplicity? It is difficult to answer this question confidently, but I believe that a scientist derives aesthetic pleasure from a theory upon noticing that it exhibits to at least a certain degree a form of simplicity for which he or she has a predilection.

This view is suggested by typical statements of scientists. For instance, in the following passage, Weinberg compares the merits of Newton's and Einstein's theories of gravitation:

> Einstein's general theory of relativity is characterized by a set of second-order differential equations; so is Newton's theory of gravity. From that point of view they are equally beautiful; in fact Newton's theory has fewer equations, so I guess in that sense it is more beautiful. But Einstein's theory of general relativity has a greater quality of inevitability. In Einstein's theory there was no way you could have avoided an inverse square law [. . .] at large distances and at slow speeds. [. . .] With Newton's theory, it would have been very easy to get any kind of inverse power you liked, so Einstein's theory is more beautiful because it has a greater sense of rigidity, of inevitability.[36]

As Weinberg explains, each of the two theories shows a particular form of simplicity to a degree greater than its competitor: Newton's theory demonstrates more parsimony with equations than Einstein's, while Einstein's theory is more parsimonious with assumptions than Newton's. In section 3 we encountered Dirac's similar comment on these two theories' simplicity properties. But Weinberg's assessment goes beyond Dirac's, since it explicitly describes a predilection for theories that show a partic-

36. Weinberg (1987), pp. 107–108. See also Weinberg (1993), pp. 106–108.

ular form of simplicity as an aesthetic preference. A scientist's having attached preference to one of these forms of simplicity ensures that he or she will regard the corresponding theory as the more beautiful. I take this as support for the conclusion that the criterion of form of simplicity is an aesthetic criterion for theory choice.

On this basis, I regard form of simplicity as a class of aesthetic properties in the sense established in Chapter 3; and I regard a particular form of simplicity, such as ontological parsimony, as an aesthetic property that theories may show. To the extent that aesthetic judgments are observer relative, judgments of theories made on the criterion of form of simplicity must also be considered observer relative.

Now let us consider whether either the criterion of degree or that of form of simplicity should be regarded as diagnostic of a theory's empirical adequacy. As we know from Chapter 5, there are two routes by which a particular criterion can be justified as promoting the choice of theories having high degrees of empirical adequacy. One route is goal analysis: logical elucidation of the notion of empirical adequacy is capable of revealing that some properties of theories are conducive to their having empirical adequacy to a high degree. The second route is inductive projection: once we have a criterion to pick out good theories, we are able inductively to identify other properties whose presence is correlated with theories' being good.

Goal analysis, I believe, sheds some light on the criterion of degree of simplicity. Some of the arguments cited in section 1 lead me to assert the following. Of two theories that differ in that they exhibit one form of simplicity to differing degrees but are equally worthy in all other respects, the theory exhibiting this form of simplicity to the greater degree is empirically superior to the other. I attribute this finding to goal analysis, as it derives from an elucidation of the notion of empirical adequacy and of how this property can be discerned in theories. In virtue of this result, I incline toward viewing the criterion of degree of simplicity as diagnostic of a theory's empirical adequacy, and therefore as an empirical criterion for theory choice.

Does goal analysis establish an equivalent result for the criterion of form of simplicity? In my opinion, goal analysis provides no reason for thinking that systematically preferring a particular form of simplicity is an effective strategy for identifying theories with higher degrees of empirical adequacy. This is because logical analysis of the notion of empirical adequacy fails to establish any conclusion of the form "The policy of choosing among theories for the possession of S_1 is more effective at yielding theories with high degrees of empirical adequacy than the policy of choosing among theories for the possession of S_2," where S_1 and

S_2 are two forms of simplicity, such as numerical simplicity, explanatory simplicity, ontological parsimony, and so on.

The second route by which one could show the empirical worth of choosing theories in virtue of a certain form of simplicity is inductive projection. One could cast an eye over the history of science and determine whether theories chosen for a certain form of simplicity have, as a matter of contingent fact, tended to show higher degrees of empirical adequacy. If such a result were established for a particular form of simplicity, one would have inductive justification for a policy of choosing theories for that form. One would then be justified in paying regard to, say, ontological parsimony and no other form of simplicity.

Establishing this result by inductive projection is an empirical task in the historiography of science. On the one hand, one can easily conceive of evidence being discerned that, say, theories chosen for ontological parsimony have in the past tended to have greater empirical adequacy than theories adopted for explanatory simplicity. This would be evidence for regarding the criterion of form of simplicity as an empirical criterion. On the other hand, the historical record seems to show no convincing correlation of the required kind between particular forms of simplicity and empirical success. If no such correlations ever appear, theory evaluations based on the form of theories' simplicity cannot be considered empirical.

Lastly, what are the implications of these findings for scientists' actual simplicity considerations, which involve both appraisals of form and of degree of simplicity? If my analysis in section 3 was accurate, the following holds. For any pair of theories that have unequal empirical adequacy, there exists a form of simplicity that the theory with the greater empirical adequacy shows to a higher degree. Our empirical interests would be best served if, in each case of theory choice, we attached preference to the form of simplicity that the empirically superior theory shows to the higher degree: for then we would be likelier to choose in each case the empirically superior theory. If this form of simplicity were the same in every case of theory choice, we would doubtless come to know which form this was, because a correlation would become apparent between a theory's having a higher degree of this form of simplicity and its demonstrating greater empirical adequacy. In this case, scientists could use the aesthetic induction to tune their aesthetic preferences to the correct form of simplicity, and they would develop the power thereby to diagnose empirical adequacy in theories.

On the other hand, as we have seen, it is possible that the aesthetic induction will never reveal any lasting correlation between a theory's having a particular form of simplicity to high degree and its demonstrating greater empirical adequacy than its competitors. In this case, it remains true that, for any pair of theories that have unequal empirical adequacy, there exists a form of simplicity that the theory with the greater empirical adequacy shows to a higher degree than does the other theory. But the form of simplicity for which this statement holds will vary from case to case, and we will not know for which form it holds unless we already have knowledge of the two theories' degrees of empirical adequacy. Where competing theories differ in their degrees and forms of simplicity, we therefore will not be able to use a criterion of form of simplicity to reveal which theory has the greatest empirical adequacy. Of course, such a criterion is still likely to play a part in determining our preferences among theories, but we will be unable to justify it on the grounds that it reveals the empirical adequacy of theories.

Revolution as Aesthetic Rupture

Certain methods have frequently yielded the most beautiful results, and many persons have been tempted to believe that the development of science to the end of all time would consist in the systematic and unremitting application of them. But suddenly they begin to show indications of impotency, and all efforts are then bent upon discovering new and antagonistic methods. Then there usually arises a conflict between the adherents of the old method and those of the new. The point of view of the former is characterised by its opponents as antiquated and obsolete; whilst its upholders in their turn look down with scorn upon the innovators as perverters of true classical science.

—Ludwig Boltzmann, "The Recent Development
of Method in Theoretical Physics"

1. THE OCCURRENCE OF SCIENTIFIC REVOLUTIONS

The model of scientific practice that I have been constructing claims that the set of criteria that scientists use in theory choice changes with time. So far, I have envisaged this change amounting only to gradual evolution: I have suggested that scientists' empirical criteria remain substantially unchanged through the history of science, and I have argued that the aesthetic induction updates aesthetic criteria gradually and continuously.

But the claim that scientists' criteria for theory choice show only gradual evolution conflicts with a large body of evidence from the history of science. It is true that, on most occasions when a scientist discards one theory for another, the discarded theory and the theory that takes its place share basic features. On other occasions, however, scientists adopt theories that differ radically from their immediate predecessors. Such an occasion arose in the early twentieth century, when classical theories of submicroscopic phenomena were replaced by quantum theories. We call these occasions scientific revolutions.[1] Both the theory that is discarded in a revolution and the theory that takes its place were presumably recommended as the best theory available at the time of their adoption by the criteria for theory choice that the community then applied. This means that, in revolutions, scientists' criteria for theory choice must undergo radical and rapid change. For this reason, the model of scientific practice that I have been constructing cannot yet be regarded as complete. In this chapter, I show how a simple extension enables the model to account for scientific revolutions as well as gradual change.

The occurrence of revolutions imposes, I suggest, four obligations on models of scientific practice. If it does not meet these, a model does not properly account for revolutions and is therefore unsatisfactory. The obligations are the following:

1. Models of scientific practice must acknowledge that science experiences both revolutions and periods during which criteria for theory evaluation remain unchanged. An account of only the revolutions or of only the latter periods does not constitute an acceptable model of scientific practice in its entirety.

2. It is not enough for a model of scientific practice to offer separately an account of revolutions and an account of the periods in which criteria remain unchanged: these accounts must be causally connected to one another. We wish to know how one mode of development follows upon the other, and especially what factors trigger and terminate revolutions.

3. Models of scientific practice must acknowledge that a scientific revolution constitutes a radical transformation in a community's criteria for theory choice. For example, in the revolution accompanying the rise of quantum theory, the community switched from insisting that theories be visualizing and deterministic to adopting theories that were abstract and indeterministic.

4. On the other hand, revolutions must not be portrayed as being so deep that they leave no element of scientific practice unchanged. We would be unable to describe a discipline as undergoing a revolution unless its postrevolutionary form could be identified as a continuation of its prerevolutionary form.

Let us investigate how well these obligations are met by one of the models of scientific practice now in existence. Accounts of scientific revolution have been proposed at least since the 1930s, when Gaston Bachelard described science as undergoing *ruptures épistémologiques* and Ludwik Fleck spoke of science's successively adopting different *Denk-*

---

1. The historical evidence that science undergoes revolutions is surveyed by I. B. Cohen (1985), pp. 40–47, 389–404.

stile.[2] But today the most influential account of scientific revolutions is that of Kuhn. He presents history as sectioned into periods of normal science, each of which is characterized by a paradigm and terminated by a revolution.

Kuhn's model of scientific practice fully meets obligation 1: indeed, an insistence that scientific practice contains both periods of continuity and revolutions is one of his model's most original features. Obligation 2 is satisfied less well: Kuhn's model does not specify clearly by what factors a revolution is triggered and terminated. Indeed, Lakatos criticized Kuhn for citing nothing more definite than mob psychology to account for revolutions.[3] But the most serious flaw of Kuhn's model is that it is unable to meet both obligation 3 and obligation 4. How this inability manifests itself depends on where the emphasis is placed within Kuhn's publications.

In some of his best-known passages, Kuhn claims that there are no criteria for theory evaluation that scientists in different paradigms share. The conceptual resources provided by two successive paradigms are so unlike that their members "live in different worlds."[4] On this reading, Kuhn's model has no difficulty in portraying scientists as working in different styles at different times: however, it fragments the history of science into periods that share nothing, and therefore fails to meet obligation 4. Other passages of Kuhn's show more moderation. There are, he says, five "good reasons for theory choice" that are common to members of all paradigms: the criteria of accuracy, consistency, simplicity, breadth of scope, and fruitfulness.[5] In Kuhn's statement that "it is vitally important that scientists be taught to value these characteristics" of theories,[6] the historically indiscriminate reference to scientists makes sense only on the assumption that these criteria are justified in all paradigms. Unless Kuhn can identify some further category of criteria for theory choice that are paradigm specific, it remains unclear how any deep transformations in scientific practice can occur. On this reading, Kuhn's model no longer satisfies obligation 3.

This weakness of Kuhn's model derives, I think, from an assumption that all precepts that scientists follow—and, in particular, all criteria for theory evaluation—can be treated adequately as belonging to one set. Kuhn does not draw distinctions among, for instance, the five values for

2. Bachelard (1934), especially pp. 50–55; Fleck (1935), especially pp. 125–145. For a history of models of scientific revolution, see I. B. Cohen (1985).

3. Lakatos (1970), p. 178.

4. Kuhn (1962), pp. 111–135.

5. Kuhn (1970), p. 261; Kuhn (1977), pp. 321–322. See also Kuhn (1962), pp. 144–155.

6. Kuhn (1970), p. 261.

Revolution as Aesthetic Rupture

theory assessment that he lists. He claims in his moderate writings that these values all endure across revolutions, and in his radical writings that a revolution overthrows them all: in either case, every precept exhibits identical behavior. A more promising route, I suggest, would be to portray one group of precepts as liable to radical changes, which would account for the occurrence of revolutions, and to attribute transparadigmatic validity to some other group, which would assure the continuity of scientific practice through revolutions.

The model of scientific practice that I have been developing is well placed to follow this route. From the outset we have identified two sets of criteria for theory evaluation, which have different origins and show different behavior. One is the set of empirical criteria, which are formulated by goal analysis and show little change in time. The other is the set of aesthetic criteria, which originate in inductive projection and alter in response to the perceived performance of past scientific theories. We shall now discover how the evolution of scientists' aesthetic criteria leads, in certain circumstances, to revolution.

2. THE ABANDONMENT OF AESTHETIC COMMITMENTS

As we saw in Chapter 5, the aesthetic induction ensures that scientists' aesthetic criteria are conservative: in cases of theory choice, they attribute most value to, and recommend for adoption, theories that have the aesthetic properties shown by the empirically most successful theories adopted previously. Let us consider what effect this conservatism has on scientists' ability to choose at each time the empirically most successful theory available.

In a particular state of affairs, a community's aesthetic canon will not deter it from adopting the empirically best-performing theories on offer. This state of affairs endures as long as the new theories that become available maintain the correlation manifested in past theories between having particular aesthetic properties and achieving empirical success. Let us examine why this condition ensures that the community's aesthetic canon agrees with its empirical criteria about which theories should be adopted. As before, E_P is the degree of empirical adequacy that a community attributes to the set of its theories that possess aesthetic property P, and W_P is the weighting given to P in the community's aesthetic canon. If a community is accustomed to seeing great empirical success being demonstrated by P-bearing theories, the values of E_P and, thanks to the aesthetic induction, of W_P will be high. Now, if new theories that become available are either empirically successful and show P, or are unsuccessful and lack P, the values of E_P and W_P will remain unchanged. This means that the recommendations for theory choice issued by the aesthetic canon will also remain unaltered. So if the community's subsequent empirically successful theories persist in showing particular aesthetic properties, they will find an aesthetic canon apt to value them highly.

In the state of affairs that we are imagining, each new empirically successful theory that arises within the community differs in aesthetic properties to only a small extent from previous successful theories. As long as this holds true, the aesthetic canon is able to evolve fast enough to maintain pace with the evolution of the aesthetic properties exhibited by the sequence of empirically successful theories. Then, the aesthetic properties of the empirically most successful theory available to the community at each time will win favor with the aesthetic canon.

In this state of affairs, theory choice is uncontroversial: at least, there are no controversies caused by scientists' being compelled to weigh the aesthetic appeal of some theories against the observational success of others. This phase of science corresponds to a period of what Kuhn calls normal science, which he similarly regards as marked by consensus in theory choice.[7] What Kuhn calls a paradigm corresponds on this account to the aesthetic canon that contributes to determining theory choice during this period. As we saw in Chapter 5, if there is a long run of empirically successful theories that have similar aesthetic properties, an aesthetic canon may become deeply entrenched and therefore maintain the ascendancy over a community that Kuhn attributes to paradigms. The staple problems of normal science are, for Kuhn, "puzzles" that are solved in the manner that the paradigm prescribes.[8] On my view, puzzles are problems that are solved by theories that accord with the aesthetic canon in force. On this view, as on Kuhn's, while such solutions may be difficult to construct, their acceptability is generally not disputed: it is the essence of such contributions that they accord fully with the stipulations of the canon for theory choice.

Scientific practice assumes this placid character only as long as the new theories that become available maintain the correlation manifested in past theories between having particular aesthetic properties and achieving empirical success. If new theories do not show these correlations, the recommendations of the aesthetic canon about which theories to adopt will depart from those of empirical criteria. Once more, imagine a community that is accustomed to seeing P-bearing theories demon-

---

7. For Kuhn's characterization of normal science see Kuhn (1962), pp. 23–34.
8. The procedures of puzzle solving are discussed by Kuhn ibid., pp. 35–42.

Revolution as Aesthetic Rupture

strate empirical success and that therefore attributes high values to E_P and W_P. A worsening of the empirical performance of available P-bearing theories is reflected in E_P and W_P only after a certain time lag. Because of this, changes in the recommendations of the aesthetic canon will lag behind developments in the empirical capability of available theories. The aesthetic canon will continue to express preference for aesthetic properties that were exhibited by the community's former best theories but that are not shown by its current best theories: the advice of the community's aesthetic canon in theory choice will depart from that of empirical criteria.

Scientists will probably experience this development in the following way. As long as a community remains able to solve any problems thrown up in research by means of theories that are endorsed by both its aesthetic canon and its empirical criteria for theory choice, no dilemmas arise. In time, however, the community encounters harder problems, which appear not to admit solutions acceptable on both aesthetic and empirical criteria: the proposed solutions that best satisfy the aesthetic canon demonstrate less empirical success than some solutions that violate it. These problems correspond to those that Kuhn calls anomalies.[9] At first, scientists may ignore this conflict between the two sets of criteria for theory choice: but it will eventually demand a principled resolution. Each scientist can resolve it in his or her own mind by designating one of the sets of criteria as overriding the other and adopting whichever theories this set recommends. But nothing ensures that the same set of criteria will be selected by all members of the community as overriding the other. Two options are available, and each will be pursued by a group of scientists.

One group, which I shall call the conservative faction, will designate the aesthetic canon as overriding the set of empirical criteria. This option has what conservative scientists will see as a great virtue: it ensures that the theories that they are led to adopt have the aesthetic properties that they have become used to seeing in empirically successful theories. Conversely, since it consists in weakening empirical concerns in theory choice, this option will generally lead scientists to adopt theories that are empirically less successful than some that are available. However, members of the conservative faction may reason away this apparent disadvantage by some of the arguments that we examined in Chapter 6: they may suggest that theories that exhibit what they see as beauty are bound to be closer to the truth than theories that lack beauty, even if the latter accord better with available empirical data.

9. Ibid., pp. 66–76.

The other group, which I shall call the progressive faction, will designate empirical criteria as overriding the aesthetic canon. This is the option that would be recommended by Willard V. O. Quine, who says that
it is permissible for scientists to pursue elegance in theories "as long as
it is appealed to only in choices where the pragmatic standard prescribes
no contrary decision."[10] Since this option amounts to relaxing extraempirical constraints on theory choice, it will permit the progressive faction to adopt theories empirically more successful than those adopted by
their conservative colleagues. However, on the established aesthetic
canon, these theories would be judged less pleasing that those of the
conservatives. In short, like the Cavaliers and the Roundheads in the English Civil War in 1066 and All That, the theories adopted by the conservative faction are Wrong but Wromantic, while the ones adopted by the
progressive faction are Right but Repulsive.[11]

The fact that the theories adopted by the progressive faction are
viewed as displeasing by the established aesthetic canon will certainly
be cited by conservative scientists as evidence that their progressive colleagues are taking the wrong road. Some members of the progressive
faction might find this fact troubling as well: after all, they too will have
had a great commitment to their discipline's aesthetic canon. To neutral-
ize this worry, members of the progressive faction may profess indiffer-
ence toward all aesthetic properties of theories. To justify this stance, they
may reason that the community's previous aesthetic commitments have
only hampered the pursuit of empirical success. The progressive faction
can declare that, now that aesthetic constraints on theory choice have
been relaxed, the community has become free to make the empirically
most fruitful choices among competing theories.

This, I suggest, is how a scientific revolution should be interpreted: as
the repudiation of aesthetic constraints that a community had become
accustomed to imposing on theory choice. I see the progressive faction's
abandonment of the established aesthetic canon, and their resolution to
conduct theory choice unhampered by aesthetic commitments and in the
pursuit exclusively of empirical success, as the revolutionary act. As we
expect of a revolutionary act, it consists of a disavowal by some community members of commitments previously accepted by the community in
its entirety.[12]

Two factions such as I have described formed in the physics commu-

10. Quine (1953), p. 79.
11. Sellar and Yeatman (1930), p. 63.
12. I first advanced the view of scientific revolutions as aesthetic ruptures in McAllister (1989), pp. 41–47.

nity during the rise of quantum theory. As we shall see in Chapter 11, the conservative faction in this revolution—which included Planck and Einstein—decided that the aesthetic canon that had been built up by classical physics should override the standard empirical considerations in theory choice. This policy prevented them from endorsing what had become the empirically best-performing theory of subatomic phenomena on offer: quantum theory. Einstein defended this constraint by arguing that such an aesthetically unappealing theory was certain to be far from the truth, whatever its current empirical performance was. The progressive faction, led by Niels Bohr, subordinated the aesthetic canon to empirical criteria. In response to Einstein's aesthetics-based criticism of quantum theory, Bohr suspended allegiance to any aesthetic preferences among theories and embraced a form of positivism in theory choice.

The conservative and progressive factions may coexist in the community for a while. They will continue to adopt different theories, however: the progressive faction will continue to choose among theories on empirical criteria, while the conservative faction will persist in preferring theories showing the familiar aesthetic properties. The gap in empirical performance between the theories adopted by the two factions will therefore continue to grow. One imagines that, once this gap has reached some great magnitude, members of the conservative faction will come to see the progressive faction's theories as preferable to their own, notwithstanding their unappealing aesthetic properties. The conservative faction will gradually relax its commitment to their aesthetic canon. When the entire community has aligned itself on the progressive faction's policy of theory choice, the community's divisions are overcome and the revolutionary phase is terminated.

The effect of the revolution has been to strip the community of one of its two sets of criteria for theory assessment: as the revolution progresses, the community loses its commitment to an aesthetic canon. This means that theories that do not conform to the old aesthetic canon will, relatively suddenly, cease to encounter opposition. The sudden collapse of the opposition to new aesthetic forms occurs also in art. As an established aesthetic canon loses its grip on an artistic community, some innovative works of art may suddenly become much more acceptable than before. This parallel between the sciences and the arts explains how the physicist John A. Wheeler could meaningfully extend to present-day physics a remark that Gertrude Stein made of modern art. Stein described the change in the perception of an innovative artwork as follows: "It looks strange and it looks strange and it looks very strange; and then suddenly it doesn't look strange at all and you can't understand what made it look strange in the first place."[13] Wheeler believes that scientists' opinions of innovative theories change in the same way. This change is easily explained: it ensues from the community's disavowal of an aesthetic canon that regarded the theory unfavorably.

In the aftermath of a revolution, some empirically minded scientists may think that the community's abandonment of its aesthetic commitments has changed the course of science irrevocably. They may hope that the community will never again impose aesthetic constraints on theory choice and that it will henceforth select among theories exclusively with the aim of obtaining the best possible empirical performance. They may imagine science becoming, perhaps for the first time, a quest for empirical success free of extraempirical concerns. I suggest that these scientists will invariably be disappointed. Once a revolution is over, the aesthetic induction will return to affect scientists' preferences. Scientists will begin again to discern correlations between some aesthetic properties of theories and high degrees of empirical success. They will come to believe that theories are assured of empirical success if they show these aesthetic properties. These properties will be attributed a weighting in an aesthetic canon on the basis of which the community will begin to conduct theory choice. In time, of course, divergences will appear between the recommendations of this canon and empirical criteria, until a new crisis in theory choice and eventually a further scientific revolution occur.

3. Continuity and Rupture in Revolutions

Let us examine how the model of revolution that I have formulated satisfies the obligations that I set out in section 1. First, there is no doubt that this model acknowledges that science experiences both revolutions and periods during which criteria for theory evaluation remain unchanged. The latter periods are those in which a community's theory choice is dominated by a particular aesthetic style; revolutions are episodes in which a community's aesthetic canon is overthrown. Second, the model explains how one mode of development follows upon the other: a revolution is triggered when the aesthetic constraints that a community imposes on theory choice begin too severely to hamper the pursuit of empirical success. Lastly, obligations 3 and 4 prescribe that models of scientific practice should allow that revolutions are deep ruptures in theory choice that nonetheless leave aspects of scientific practice unaltered.

13. Wheeler (1983), p. 185.

These obligations too are fully satisfied by the model presented here, as we shall now see.

The characteristic feature of my model is that it interprets revolutions as changes in no more than a subset of the criteria on which communities perform theory choice: although in a revolution one aesthetic canon eventually replaces another, empirical criteria survive unchanged. On the one hand, therefore, this model portrays a revolution as a transformation so deep as to change a community's style of theorizing. On the other hand, there are some criteria that scientists before and after a revolution share.

On this score, we may compare Kuhn's model of revolutions and my own. As we have seen, Kuhn suggests on some occasions that all scientists share five criteria for theory choice, and on other occasions that there exist no paradigm-independent criteria for judging rival theories. Therefore, Kuhn and I disagree over how fully scientists can communicate their reasons for theory choices across a revolutionary divide. Kuhn claims either that they would understand each other's reasons fully or that they would find them entirely incomprehensible. I claim that they would regard each other's aesthetic preferences as alien but would recognize each other's understanding of and concern for empirical properties of theories, such as internal consistency or consistency with empirical data. Which of these claims accords more closely with evidence from scientific practice?

An example of a discussion about the merits of a theory that was conducted across a revolutionary divide is the debate between Bohr and Einstein about quantum theory.[14] In this debate, Einstein is the member of the conservative faction who maintains the established aesthetic commitments, while Bohr is the member of the progressive faction who has repudiated all aesthetic commitments. What is most notable about this debate is that Bohr and Einstein show only partial incomprehension of each other's reasons for choosing among theories. They show incomprehension when they discuss whether there are any metaphysical and other aesthetic properties that a theory must have in order to be acceptable. Bohr does not comprehend the rationale for Einstein's insistence that no indeterministic theory is satisfactory, and Einstein does not understand how Bohr can be content with a theory as unappealing as quantum theory. In contrast, they show full understanding of each other's appeals to empirical criteria: for instance, they agree over what it is for a

14. Among the many studies of the Bohr–Einstein debate are Jammer (1974), pp. 109–158, Honner (1987), pp. 108–141, Murdoch (1987), pp. 155–178, and Kaiser (1994); the last of these contains further references. For Bohr's own account see Bohr (1949).

theory to accord with a body of empirical data and to be internally consistent. The long section of their discussion about whether quantum theory is internally inconsistent, for instance, shows them disagreeing about matters of fact, but not talking past one another. Episodes such as these, I suggest, show that what limits communication between members of different paradigms is not an all-pervading incommensurability but rather a partial lack of commonality owed to the difference between the aesthetic canons that predominate at different times.

4. UNDERSTANDING PAST SCIENCE

The model of scientific revolutions that I have presented has interesting implications for historiography of science. If Kuhn's more radical claims about scientific revolutions were accurate, historiography of science would be far more difficult than many of its practitioners imagine. If it were true that a revolution effects a complete change in criteria for theory choice, a factor that constituted a reason for preferring one theory to another in a given paradigm would not do so in a different paradigm. This means that even the best arguments offered by scientists to justify their choices among theories would not be convincing to a historian studying the episode after a revolution. The historian would thus be unable to make sense of most past episodes of theory choice.

Historiography of science is less difficult according to the model of scientific practice that I have presented. Admittedly, our understanding of scientists' theory choices in periods previous to our own will still be imperfect. After all, we do not share the aesthetic canons that characterized those periods and contributed to deciding the theory choices performed within them. This means that the aesthetic reasons that were adduced in these periods as grounds for preferring one theory to another will fail to convince us. For example, we are not persuaded by aesthetic arguments against Kepler's theory of the solar system, because we do not share the sixteenth-century commitment to describe heavenly motions as combinations of circles; we do not find Einstein's opposition to quantum theory warranted, because we no longer insist that physical theories should be deterministic. However, empirical criteria endure largely unchanged over time. In consequence, what constitutes an empirical reason for preferring one theory to another in a given period will retain its force in later periods. Hence, at least a part of the overall grounds that scientists adduce for preferring one theory to another will remain comprehensible to later historians. My model thus portrays theory choices of the past as more intelligible to historians than Kuhn's model does.

The experience of historians of science supports my model. Let us take as an example the grounds for theory choice in sixteenth-century planetary astronomy. Without doubt at least one revolution in this discipline has occurred since then, so we will find ourselves here grappling with changes in criteria for theory choice. Some of the reasons adduced in favor of theories in sixteenth-century planetary astronomy strike us as cogent, but others do not. On the one hand, as Keith Hutchison has documented, astronomers were accustomed to attaching value to theories partly to the extent that they suggested analogies between celestial structures and political institutions.[15] For instance, Copernicus argued that a virtue of his heliocentric theory was the fact that it portrayed the solar system as analogous to a court:

> At rest [. . .] in the middle of everything is the sun. For in this most beautiful temple, who would place this lamp in another or better position than that from which it can light up the whole thing at the same time? For, the sun is not inappropriately called by some people the lantern of the universe, its mind by others, and its ruler by still others. [. . .] Thus indeed, as though on a royal throne, the sun governs the family of planets revolving around it. Moreover, the earth is not deprived of the moon's attendance. On the contrary, as Aristotle says, the moon has the closest kinship with the earth.[16]

Present-day astronomers see no justification for valuing theories on the strength of analogies between celestial structures and political institutions. This, I suggest, is because since the time of Copernicus there has been a change in astronomers' aesthetic canons for theory choice. On the other hand, sixteenth-century astronomers valued theories for such logical and empirical properties as being internally consistent and according with empirical data, and these properties are still seen as strengths of theories today. For instance, the reconstruction of Copernicus's theory by Noel M. Swerdlow and Otto Neugebauer attributes persuasive force to logical and empirical properties that Copernicus too would have claimed as its virtues.[17] This is because a rationale has been found for these logical and empirical properties both in the sixteenth century and in the present day. As this example illustrates, historiography of science supports my suggestion that scientific revolutions consist of a change in aesthetic canons rather than of a wholesale substitution of criteria for theory choice.

15. Hutchison (1987), pp. 97–109.
16. Copernicus (1543), p. 22.
17. Swerdlow and Neugebauer (1984).

5. Factors Inducing and Inhibiting Revolutions

Kuhn's model of revolutions and mine can usefully be compared also for the claims that they make about the factors that induce and inhibit scientific revolutions.[18]

In Kuhn's view, scientists choose among theories in the light of their empirical and aesthetic properties, but these sets of properties play different roles during periods of normal science and during revolutions. According to Kuhn, aesthetic factors play no decisive role in theory choice within normal science. In the puzzle solving of which normal science consists, he says, the usual stimulus for a scientist to adopt a new theory is its being demonstrated empirically superior to its competitors. In a revolution, by contrast, empirical considerations will typically weigh in favor of their current paradigm and against paradigm switch. After all, Kuhn says, a mature paradigm will have developed a track record in problem solving that cannot be matched by a new paradigm.[19]

Kuhn identifies the factors that tend to induce paradigm switch in arguments of a different sort: "These are the arguments, rarely made entirely explicit, that appeal to the individual's sense of the appropriate or the aesthetic—the new theory is said to be 'neater', 'more suitable', or 'simpler' than the old."[20] Kuhn suggests that, but for such arguments, a new paradigm might never be adopted: "The importance of aesthetic considerations can sometimes be decisive. Though they often attract only a few scientists to a new theory, it is upon those few that its ultimate triumph may depend. If they had not quickly taken it up for highly individual reasons, the new candidate for paradigm might never have been sufficiently developed to attract the allegiance of the scientific community as a whole."[21]

This means that, in a revolutionary crisis, empirical and aesthetic considerations are aligned on opposite sides. Empirical grounds militate in favor of maintaining the established paradigm but may be outweighed by aesthetic considerations:

> Something must make at least a few scientists feel that the new proposal is on the right track, and sometimes it is only personal and inarticulate aesthetic considerations that can do that. Men have been converted by them at times when most of the articulable technical arguments pointed the other way. When first introduced, neither Copernicus' astronomical

18. I compare my model of revolutions with Kuhn's also in McAllister (1996).
19. Kuhn (1962), pp. 156–157.
20. Ibid., p. 155.
21. Ibid., p. 156.

Revolution as Aesthetic Rupture

theory nor De Broglie's theory of matter had many other significant grounds of appeal.[22]

Kuhn's expectations about the role of empirical and aesthetic factors in inducing and inhibiting revolutions are thus the converse of mine. My model predicts that, in a choice between a theory of familiar aesthetic form and one showing radically new aesthetic properties, scientists' aesthetic preferences will weigh in favor of the former. If the theory showing new aesthetic properties is ever accepted, it will be because its empirical performance is good enough to outweigh the aesthetic dislike of it that scientists will initially feel. I thus see the aesthetic properties of revolutionary theories as factors tending to inhibit revolutions, and their empirical performance—if it is good enough—as the factor tending to induce them.

These predictions may be tested against historical evidence. To this end, we must identify a theory whose adoption marked a revolution in some branch of science. We must then ascertain what role the empirical and aesthetic properties of that theory and its displaced predecessor played in inducing or inhibiting the paradigm switch. If it were found that the paradigm switch had been inhibited by empirical considerations and induced by aesthetic factors, Kuhn's model of revolutions would be corroborated. If the evidence showed the opposite, mine would be.

I will conduct this test in Chapters 10 and 11. For now, I shall offer only a conjecture about why Kuhn should regard aesthetic factors as those that induce revolutions. Kuhn appears to regard scientists' aesthetic preferences as highly idiosyncratic: he speaks of "the individual's sense of [. . .] the aesthetic," of aesthetic considerations as "highly individual reasons" for which to accept a paradigm, of aesthetic considerations as "subjective," "personal," even "mystical."[23] If aesthetic preferences were so idiosyncratic, then presumably at any time scientists would find theories of many different sorts aesthetically attractive, only a few of which would be represented within any paradigm. Thus, aesthetic preferences would dispose scientists to switch from established theories to theories showing new aesthetic properties.

In reality, while a community's aesthetic canon changes with time, the aesthetic preferences of scientists at any one time do not diverge very strongly: there is wide agreement about the aesthetic properties that theories should possess. This is because, far from being conceived at whim,

22. Ibid., p. 158.
23. Ibid., pp. 155, 156, 158. Kuhn's assumption that any aesthetic factors affecting paradigm choice must be nonrational is contested by Machan (1977).

scientists' aesthetic preferences are formed in a communitywide induction over the empirical performance of past theories. Of course, if aesthetic preferences are indeed formed in this manner, they are more likely to reinforce the allegiance of scientists to established theories than to induce revolutions.

6. THE ANALOGY WITH MORAL AND POLITICAL REVOLUTIONS

Since the mid–seventeenth century, parallels have been drawn between revolutions in science and revolutions in society.[24] So far, these parallels have been impaired by the unavailability of a determinate model of scientific revolutions. The model that I have presented enables deeper parallels to be identified.

Aesthetic canons in science develop in the same way in which moral codes develop in society. Moral codes prescribe patterns of behavior that have previously proved fruitful in a community: they evolve, but more slowly than new patterns of behavior arise. In times of social stability, the evolution of moral codes may be fast enough to accommodate all prevalent behavior, but this rate of evolution will be perceived as constrictive in times of change. Some people will acquire interests whose pursuit demands behavior that conflicts with the current moral code. Conservatives will refrain from such behavior out of allegiance to the code, but others will be willing to violate the code in order to further their interests. Such people will portray themselves only as abandoning outdated conventions, but conservatives may regard them as immoral or anarchistic. Despite conservative disapproval, the benefits of the new patterns of behavior may persuade others to relax their adherence to the code. Once these new patterns have become entrenched, a new morality will develop that sanctions them.

An even deeper parallel holds between my model of scientific practice and the Marxist model of the development of societies, known as historical materialism. According to Marxist theory, a society's productive capabilities exert a strong influence on its mode of organization. Developments in productive capabilities engender changes in organization. If a society relies for a sufficiently long time on particular productive capabilities, it has an opportunity of developing a mode of organization that permits their full and efficient exploitation. But new productive capabili-

24. For notes on the history of the analogy between political and scientific revolutions, see I. B. Cohen (1985), pp. 7–14, 473–477, passim; Feuer (1974), pp. 252–268, explores some of their disanalogies.

ties will eventually be developed. Although most modes of organization are sufficiently flexible to accommodate some advance, after some time the productive capabilities will have developed so far that their full exploitation is impossible within the established mode of organization. This conflict is resolved in a revolution in which the established, counterproductive mode of organization is overthrown and replaced by one better attuned to the new productive capabilities. Thus, productive capabilities successively engender, consolidate, come to be hampered by, and ultimately overthrow and replace the society's mode of organization.[25]

The analog of productive capabilities is, in science, the empirical capabilities of theories. A lineage of theories that a scientific community adopts accommodates a superstructure consisting of an aesthetic canon. As long as established theories continue to demonstrate empirical success, the aesthetic canon that reflects their aesthetic properties will become increasingly deeply entrenched. Eventually, however, a community may discover new theories that show good empirical performance but do not conform to the established aesthetic canon. Since it weighs against the adoption of these new theories, the aesthetic canon now hampers the community's empirical progress. The tension between what is desirable on empirical grounds and what is admitted on aesthetic criteria will undermine and ultimately destroy the established aesthetic canon.

25. For an account of the Marxist theory of history, see G. A. Cohen (1978).

Induction and Revolution

in the Applied Arts

1. AESTHETIC JUDGMENTS AND UTILITARIAN PERFORMANCE

I have argued that one of the sets of criteria that scientists use in theory choice is a canon produced by an inductive mechanism in which certain properties of scientific theories are weighted according to previous empirical performance: the aesthetic canon. In calling this canon aesthetic, I ascribe to it all the standard connotations of the term, such as a sensuous dimension and a connection with aesthetic values such as beauty.

We have already met two reasons for considering this canon for theory choice as genuinely aesthetic. Firstly, when scientists judge theories for such properties, they customarily use terms of aesthetic appreciation, such as "beautiful," "elegant," or "ugly." Interpreting the properties of symmetry, simplicity, and so on as aesthetic has the advantage of allowing us to take these expressions at face value. Secondly, some of the properties of theories to which value is attached by the canon, such as symmetry, simplicity, reliance on analogy, and visualization, are archetypally aesthetic: in virtue of possessing such properties, objects are liable to strike beholders as having a high degree of aptness. These reasons are enshrined in the criteria that I formulated in Chapter 2 for recognizing which properties of theories are aesthetic.

However, the reader might still be skeptical that what is produced by an inductive projection over the empirical performance of theories can truly be considered an aesthetic canon, believing that judgments that are truly aesthetic are unrelated to considerations of empirical performance or utility. I aim to allay this skepticism in the present chapter. By investigating how the exploitation of new materials in architecture and industrial design gives rise to new aesthetic canons, I shall suggest that aesthetic canons in the applied arts respond to utilitarian performance in exactly the way in which aesthetic canons in science do. We begin by examining the impact that cast and wrought iron and steel had on architectural design in Britain and France.[1]

2. The Response of Architectural Design to Iron and Steel

In Britain, cast and wrought iron has been used since the seventeenth century in domestic and decorative fittings, such as firebacks and railings. From the early eighteenth century, cast iron was employed occasionally also in a structural capacity: for instance, Christopher Wren used cast-iron chains to counteract the outward thrust of the brickwork in the dome of St. Paul's Cathedral (1675–1710) and cast-iron columns to support galleries in St. Stephen's Chapel in the Palace of Westminster (1714), where the House of Commons then sat.[2] However, these features did not disrupt established design principles: the iron chains in St. Paul's are hidden from view, and the columns in St. Stephen's Chapel were treated as internal fittings rather than structural elements. Iron began to affect design at the end of the eighteenth century, when it enabled certain needs to be met in bridges and industrial buildings.[3]

The customary material for bridges had long been masonry. However, toward the end of the eighteenth century, ironmasters and engineers who had gained familiarity with cast iron realized that it permitted the construction of bridges with relatively long spans. The ironmaster John Wilkinson recommended the use of iron when plans were drawn up to bridge the River Severn at Coalbrookdale in Shropshire, the county at the center of pioneering work in iron casting: his efforts resulted in 1779 in the world's first cast-iron bridge, designed by Wilkinson and the architect Thomas F. Pritchard and built by the ironmaster Abraham Darby.[4]

1. This chapter is developed from McAllister (1995). A survey of the evolution of materials of construction is given by Elliott (1992), of which pp. 67–108 discuss iron and steel and pp. 165–197 reinforced concrete. On the effect of materials on aesthetic canons in architecture, see Guedes (1979), Mark and Billington (1989), and Pawley (1990), especially pp. 69–94, 140–161.
2. On the use of cast iron in St. Paul's Cathedral and St. Stephen's Chapel, see Strike (1991), pp. 9–13.
3. Useful discussions of the use of cast iron in a structural capacity in architecture are Giedion (1941), pp. 163–290, Gloag and Bridgwater (1948), pp. 53–236, Pevsner (1960), pp. 118–140, Pevsner (1968), pp. 9–20, 147–149, and Strike (1991), pp. 6–51, 62–71.
4. On the Coalbrookdale bridge and the other early iron bridges, see Cossons and Trinder (1979).

The civil engineer Thomas Telford, who was county surveyor to Shropshire, built no fewer than five iron bridges in the county. The first of these, over the River Severn at Buildwas (1796), was of particular importance, since it contained notable improvements in design which reduced the amount of iron needed. Another engineer, John Rennie, erected several iron bridges, including one over the River Witham at Boston, Lincolnshire (1803), and the Southwark Bridge over the River Thames in London (1819). Later in the nineteenth century, several further iron bridges were built by the engineer Isambard Kingdom Brunel. As these examples illustrate, the design and construction of iron bridges were the work of civil engineers rather than architects.

Another practical need that prompted the use of iron in a structural capacity was fireproofing. Fire was a great worry in the eighteenth century wherever people congregated for work, as in factories and warehouses, or for entertainment, as in theaters. Textile mills traditionally had internal structures of heavy timber columns and beams. Since they were lit by naked flames, and the machinery that they housed used inflammable lubricants, they were very vulnerable to fire. In the last years of the eighteenth century, several mills burned down at great cost, including in 1791 the Albion Flour Mill in London. It became imperative for mill owners to find ways of making their buildings incombustible. Masonry is of course fireproof, but its great weight makes it unsuitable for buildings of many stories. Cast-iron frames were developed largely in response to these needs. Their designers were primarily not architects but, as were the designers of the early iron bridges, engineers: often the same engineers who were simultaneously using cast iron in jennies, looms, and the steam engines that powered them. The engineer William Strutt and Richard Arkwright, the inventor of the spinning jenny, erected a six-story cotton mill at Derby in 1792–1793 which had iron columns (though it still retained timber beams, protected by plaster sheathing) and was described as fireproof. Matthew Boulton and James Watt, the engineers who perfected rotary steam engines, constructed a much imitated seven-story cotton mill in Salford in 1801 which employed not only cast-iron columns but also I-section cast-iron girders to support the floors.[5]

These early cast-iron bridges and iron-framed industrial buildings had designs that were very innovative: they were shaped by the desire to exploit to the full the technical capabilities of the material. The development of these designs was facilitated by the demarcation between engineering and architecture. Engineers regarded their building tasks primarily as technical problems for which their familiar material, iron,

5. On the early iron-framed textile mills, see Skempton and Johnson (1962).

Induction and Revolution in the Applied Arts

offered the most appropriate solution. Since they were not expected to follow architectural styles and mannerisms, they were free to use iron in the designs that they believed would most fully harness its capabilities.[6] In the longer term, however, the demarcation between the professions retarded the spread of iron-inspired designs to architecture. The engineers' use of iron alerted architects to its potential to meet building needs. But the work of members of the architectural profession was governed by aesthetic canons that had been evolved before iron had become available and that drew their justification from the technical capabilities of pre-existing materials such as masonry. Many of the architects who first came to use iron in a structural capacity felt unable to follow engineers in using iron in the designs best suited to exploit it fully; instead, they felt obliged to be "architectural" and continue to apply the prevailing aesthetic guidelines.

The resistance of architects to iron-inspired designs is articulated most clearly by John Ruskin in Britain and Gottfried Semper in Germany. In 1849, while admitting that the use of iron might stimulate the development of designs appropriate to its special properties, Ruskin expresses the hope that architects will continue to regard only longer-established materials as fully architectural:

Architecture [. . .] having been, up to the beginning of the present century, practised for the most part in clay, stone, or wood, it has resulted that the sense of proportion and the laws of structure have been based [. . .] on the necessities consequent on the employment of those materials; and that the entire or principal employment of metallic framework would, therefore, be generally felt as a departure from the first principles of the art. Abstractly there appears no reason why iron should not be used as well as wood; and the time is probably near when a new system of architectural laws will be developed, adapted entirely to metallic construction. But I believe that the tendency of all present sympathy and association is to limit the idea of architecture to non-metallic work; and that not without reason.[7]

Semper maintained a similar opposition to iron designs as late as 1863. His reasoning shows the effect of familiarity with masonry: he writes that iron elements are visually displeasing because their small cross sec-

---

6. Billington (1983) regards the designs of engineers in iron and concrete as contributions to an art form distinct from architecture, "structural art." I find this view difficult to reconcile with the fact that, as we shall see, these designs were gradually incorporated into mainstream architecture.

7. Ruskin (1849), pp. 70–71.

tion is inappropriate to their great strength. It would be possible to make iron columns and girders as thick as masonry elements, but Semper realizes that this would be unjustified on functional grounds. The use of iron thus requires architects "to sacrifice either beauty or function; to combine both would be impossible."[8] On this basis, Semper declares that while iron is a suitable material for temporary buildings, stone is the only proper material for works of monumental art.[9]

Such views ensured that, while mid-nineteenth-century architects were sometimes willing to incorporate iron structures into their designs, they were also determined that these should be largely hidden from view, concealed behind façades or claddings in traditional materials and styles. For instance, Thomas Rickman and John Cragg used cast-iron columns and roof trusses in St. George's Church, Everton, in Liverpool (1812–1814). The slender columns give the interior a great airiness, which could not possibly have been achieved with stone. However, the exterior wall is constructed entirely out of stone and has a conventional neo-Gothic design: the character of the structure does not show through in the public face of the building.[10]

The difference between the attitudes of engineers and architects toward cast-iron structures is even better illustrated by mid-nineteenth-century railway stations, on which the two professions often collaborated. London's St. Pancras Station (1864) consists of a cast-iron train shed designed by the engineer William H. Barlow, of which the elegant pointed arch has the widest span that had yet been achieved (fig. 1); but this is entirely concealed from the street by the massive neo-Gothic terminal building in traditional masonry designed by George Gilbert Scott.[11] Here the difference between the concerns of the professions is manifest: the engineer reacts to the need for a large roofed area, while the architect provides a façade of conventional appearance. As John Gloag has remarked, in this period in Britain engineers were "putters-up of structures" while architects acted as "putters-on of styles."[12]

In the mid–nineteenth century, some architects and critics in Britain began to argue that the choice of a material for the structure of a building should be reflected in its external appearance. Such calls had some effect. The Crystal Palace in Hyde Park, London, designed by Joseph Paxton for the Great Exhibition of 1851, was a large pavilion consisting of a glazed iron frame (fig. 2): it was an unpretentious but unashamed demonstra-

8. Quoted from Herrmann (1984), p. 176.
9. Herrmann (1984), p. 179.
10. On St. George's Church, see Strike (1991), pp. 28–30.
11. On the reception of St. Pancras Station, see Simmons (1968), pp. 91–108.
12. Gloag (1962), p. 3.

Figure 1. William H. Barlow and George Gilbert Scott, St. Pancras Station, London (1864). The British Architectural Library, RIBA, London.

Figure 2. Joseph Paxton, Crystal Palace, London (1851). A calotype by Hugh Owen or C. M. Ferrier. The British Architectural Library, RIBA, London.

Induction and Revolution in the Applied Arts

tion of the possibilities afforded by cast and wrought iron and plate glass. Its style was straightforward: the exterior was treated in the same manner as the interior, and the technique of construction and structural principles were evident from both without and within. It could still not be said that cast iron had penetrated to the heart of the architectural profession. Paxton was a railway entrepreneur and former gardener with experience in building glasshouses rather than an architect, and the completed edifice was considered by many as a work of engineering rather than of architecture. But at least the Crystal Palace displayed the forms required if the technical capabilities of cast and wrought iron were to be exploited to the full; and many visitors to the Great Exhibition regarded these new forms as showing genuine architectural beauty.[13]

One of Paxton's intentions in designing the Crystal Palace may have been that it should demonstrate British technical prowess, complementing the engineering exhibits that the building was to house. The fact that an undisguised iron structure was thought appropriate for such a building tells us nothing—it might be objected—about the acceptance of iron designs in mainstream architecture. But the open and consequent use of cast iron soon spread to the design of ordinary civic buildings. Gardner's Store, on Jamaica Street, Glasgow, by John Baird (1855–1856), is a four-story building intended for retail premises. Its façade, which has an appearance of delightful lightness, is composed of rows of slender cast-iron columns which carry most of the load; the floor plan is consequently well lit and unencumbered.[14] In such buildings, the use of cast-iron structures by architects became both widespread and unremarkable.

Now let us turn to France, where similar developments occurred with a lag of a few decades. Here, cast and wrought iron was used mainly in decorative detailing until the nineteenth century. As in Britain, when cast iron was first used in a structural capacity, it was felt that its visible effect on design should be as small as possible. Among the earliest architect-designed buildings with an iron structure is Henri Labrouste's Bibliothèque Sainte-Geneviève, Paris (1843–1850).[15] The graceful vaulting of the library's reading room (fig. 3) could not have been achieved but by means of slender iron columns and arches: to this extent, the form of the room is determined by the new material. While the iron frame is visible from within the reading room, the library has a façade of masonry in a

13. On the Crystal Palace, see McKean (1994): Paxton's career is treated on pp. 13–15, visitors' responses to the building on pp. 28–29, and the dispute about whether the building was a work of architecture on pp. 40–44.

14. On Gardner's Store, see Strike (1991), pp. 68–70.

15. On the Bibliothèque Sainte-Geneviève, see Levine (1977), pp. 325–357, Levine (1982), especially pp. 154–164, and Van Zanten (1987), pp. 83–98.

Figure 3. Henri Labrouste, Bibliothèque Sainte-Geneviève, Paris (1843–1850). James Austin, Cambridge.

generally conventional neo-Renaissance design, which completely hides the structure from view. Labrouste's design for the main reading room of the Bibliothèque Nationale, the Salle des Imprimés (1860–1867), has the same outcome: the extremely slender and graceful iron columns supporting the vaulted ceiling make no impression on the building's stone exterior.[16] These designs signaled in France the same stage in the gradual acceptance of cast iron structures that in Britain had been marked by St. George's Church, Everton.

By this time, some critics began to argue that authenticity demanded that any material of construction should be used consequently and openly. In the 1860s, Eugène Emmanuel Viollet-le-Duc attributed the mediocrity of many current architectural projects to the fact that the forms imposed on buildings were not those most appropriate to the materials employed:

> We construct public buildings that lack style, because we attempt to ally forms bequeathed by certain traditions to requirements that no longer bear relation to those traditions. Naval architects and mechanical engineers do not, when building a steamship or a locomotive, seek to recall the forms of sailing ships of the time of Louis XIV or of harnessed stagecoaches. They obey unquestioningly the new principles that are given them and produce works that have their own character and their proper style.[17]

Viollet-le-Duc demands two things: that buildings should adopt the styles most suited to their materials, rather than mimic forms appropriate to previous epochs, and that the structure of a building should appear openly, not hidden behind a façade or by cladding. If cast iron is used in the frame of a building, for example, the style impressed on the entire building should be the one that permits the fullest exploitation of the technical capabilities of iron, and the structure ought to remain visible from the exterior, not clothed in masonry or stucco.

Toward the end of the nineteenth century, Viollet-le-Duc's call for authenticity was taken up by his profession, and architects began to use cast iron openly in buildings. A particularly influential demonstration of the uses of iron and steel was given in two structures erected for the Paris Universal Exhibition of 1889.

The first of these, the Galerie (or Palais) des Machines, by the architect Ferdinand Dutert and the engineer Victor Contamin, was an exhibition

---

16. On the Bibliothèque Nationale, see Van Zanten (1987), pp. 239–246.
17. Viollet-le-Duc (1863–1872), 1:186.

pavilion with a span of over one hundred meters, since demolished (fig. 4). In this building, steel and the forms appropriate to its use were not so much displayed as flaunted. The building's structure was constituted by a number of trusses or arches, each made up of two symmetric halves, which touched at a point along the centerline of the roof. Each truss thinned noticeably toward the ground, unlike masonry columns, which generally taper upwards. This building's design embodied distinctive architectural-aesthetic principles, permitted by the characteristics of steel and not seen in buildings designed with earlier materials in mind. For instance, there was no separation between beam and column, so that it was no longer possible to distinguish load from support. The effect that the gallery had on onlookers has been described by Christian Schädlich:

> All the aesthetic ideas associated with stone buildings have been turned on their head in one instant. With the point-like bearing surfaces for the great masses, the seemingly floating vaulting, and the transparency of the whole construction, in similar fashion to the related station halls, new aesthetic laws are postulated which, understandably, not all observers readily accept as a legitimate architectural medium. The architecture lives by its own laws of completely integrated and visibly composed iron design.[18]

In this building, in short, no style had been applied to the structure other than the one that arose naturally from the material used.

The second notable structure erected for the 1889 exhibition is, of course, the three-hundred-meter iron tower by Gustave Eiffel. At first, this was commonly considered a hideous monster. Even before its completion, the Artists' Protest of 1887, instigated by Charles Garnier, the architect of the Paris Opera House, and signed by among others the writers Guy de Maupassant and Emile Zola, requested that the tower not be preserved beyond the close of the exhibition, on the grounds of its ugliness. The Eiffel Tower must have appeared all the more iconoclastic in contrast to a structure that had been erected as recently as 1884 to serve a comparable purpose, the Washington Monument in Washington, D.C., which is a stone pillar in the likeness of the obelisks of ancient Egypt.[19]

The standard early defense of the Eiffel Tower was an enumeration of its practical uses, for instance in communications and in research in physics and meteorology. Such utilitarian justifications concede the aes-

18. Quoted from Friebe (1983), p. 94. For further information on the Galerie des Machines and its reception, see Crosnier Leconte (1989) and Durant (1994).
19. On the early reception of the Eiffel Tower, see Loyrette (1985), pp. 169–189, and Loyrette (1989).

Induction and Revolution in the Applied Arts

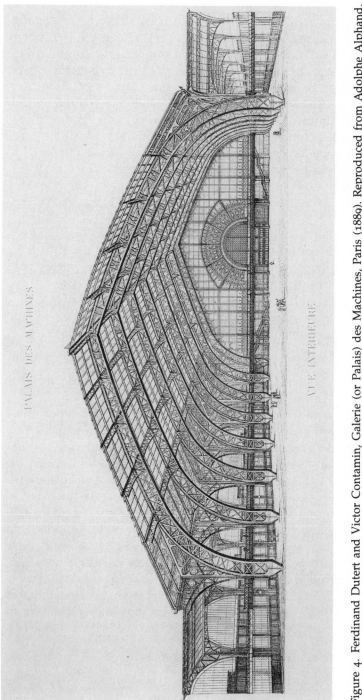

PALAIS DES MACHINES

VUE INTÉRIEURE

Figure 4. Ferdinand Dutert and Victor Contamin, Galerie (or Palais) des Machines, Paris (1889). Reproduced from Adolphe Alphand, *Exposition universelle internationale de 1889 à Paris: Palais, jardins, constructions diverses, installations générales* (Paris: J. Rothschild, 1892), série H, plate 8. The British Architectural Library, RIBA, London.

thetic ground to the critics, as if it were too much to argue that a structure like the Eiffel Tower could ever be valued for its own sake. Gradually, however, the tower began to acquire also an aesthetic defense, in virtue of the fact that the architectural aesthetic had begun to be remolded by the forms characteristic of iron. Indeed, according to J. K. Huysmans, the contrast with iron made stone appear "played out, exhausted by its repeated use" in the buildings erected for the 1889 Paris exhibition. "It could only produce better disguised or more skillfully linked borrowings from old forms."[20] By the end of the nineteenth century, the use of cast iron and steel was admitted into civic architecture in France as it had been in Britain.

In the gradual introduction and acceptance of iron-inspired designs into architecture, we may discern three phases. In the first, engineers employed iron in structures that lay outside the commonly accepted scope of architecture, such as the iron bridges and iron-framed industrial buildings of the late eighteenth century in Britain. These works acted as a demonstration of the technical capabilities of iron. In the second phase, architects were induced to use iron in a structural capacity in their own designs. However, established architectural canons, centered upon masonry, withheld aesthetic value from the new material and compelled architects to conceal iron structures behind façades or cladding.

In the third phase, concerns for authenticity encouraged a more overt use of iron. The opinion grew in strength that style ought no longer to hinder the exploitation of iron. Whereas in the earlier phases the manner of using iron would have bowed to the requirements of architectural canons, it was increasingly felt that from then on the architectural canons ought to reflect the usefulness of iron. As the critic Cornelius Gurlitt wrote in 1899: "The question [. . .] is not how to mould iron to make it conform to our taste, but the much more important one, how to mould our taste to make it conform with iron?"[21] Iron structures came to be attributed aesthetic value. An answer was found even to Semper's objection that iron structural elements were too slender in relation to their strength to give a pleasing visual impression. The architect Hermann Muthesius explained in 1913:

> One nearly always came back to the assertion that iron is too thin in order to achieve any aesthetic effect, but such an opinion presupposes that aesthetic effect can only be achieved through solidity. However,

20. Quoted from Loyrette (1985), p. 177.
21. Quoted from Gombrich (1974), p. 945. See also Gombrich's remarks, pp. 945–946, on the plasticity of taste revealed by the increasing acceptance of iron in architecture.

Induction and Revolution in the Applied Arts

such thinking involves a fallacy, whereby a traditional ideal is elevated to an absolute ideal. This traditional ideal has evolved because of the fact that until now people have built with materials that have a solid effect, namely in stone and wood; had previous generations had thin metal bars at their disposal, then probably this skeletal quality would be regarded as normal and the ideal, and solidity condemned as unaesthetic.[22]

By the conclusion of this phase, architects were no longer imposing alien styles onto iron structures but allowing iron structures to find the styles most appropriate to them. It is partly thanks to this evolution of aesthetic canons that, for example, the Eiffel Tower progressed from monster to icon.[23]

3. THE USE OF REINFORCED CONCRETE IN ARCHITECTURE

The phases in which cast iron established itself in architecture as a material with not only utilitarian benefits but also aesthetic value were negotiated a few years later by reinforced concrete.[24]

When the systematic exploitation of reinforced concrete began in the 1870s, it was used in two categories of buildings. The first was that of industrial buildings and workers' houses, where concrete was appreciated for combining practical virtues with low cost. For example, the openness of concrete frames allowed for adequate lighting in multistory factories. Its resistance to fire was also valued highly: in France, after some costly fires in the textile district of Roubaix and Tourcoing in the 1890s, several concrete spinning mills were built by the great pioneer of the material, François Hennebique. The second category of concrete buildings was that of grand mansions and public monuments. Here, concrete was used as synthetic stone, which lowered the cost of established designs. The usual practice was to apply on any exposed concrete surfaces either a cladding in masonry or a finish to mimic it. For example, in the Leland Stanford Junior Museum of Stanford University, designed by Ernest L. Ransome (1889–1891), the concrete external walls have been tooled in imitation of masonry, to complement the building's classical colonnade and other traditional features. Concrete buildings of the first

22. Quoted from Friebe (1983), p. 100.
23. On the status of the Eiffel Tower as icon, see Barthes and Martin (1964).
24. On the introduction of reinforced concrete in architecture, see Collins (1959), Pevsner (1960), pp. 179–184, Pevsner (1968), pp. 150–155, and Strike (1991), pp. 52–61, 98–116.

category exploited the technical capabilities of the material fairly fully but had little effect on architectural canons; buildings of the second category were architecturally influential but did not respond distinctively to the material's characteristics.

All the while, it was becoming evident that reinforced concrete allows forms to be created that cannot be realized with brick, stone, or iron: its plasticity allows it to assume any shape in which molds can be constructed, and its homogeneity permits traditional distinctions between building elements, such as wall and roof, to be superseded. By the end of the nineteenth century, on the strength of these observations, concerns for authenticity arose on behalf of reinforced concrete analogous to those that had been expressed a few years earlier for cast iron. Some architects and critics urged that the characteristics of reinforced concrete be allowed to dictate the manner of its presentation and decoration. Such a stance was taken in 1901 by the critic Pascal Forthuny in comments about an office and apartment block which Edouard Arnaud had built in Paris three years earlier. Fearing public disapproval of a concrete façade, Arnaud had given his building a conventional appearance by coating it with cement rendering. Forthuny expressed regret for this decision:

> Reinforced concrete is a new material, and has no links with the systems of construction which preceded it; it must thus necessarily draw from within itself its exterior aspects, which must be clearly differentiated from familiar modellings in wood, marble or stone. How can one innovate lines and surface modellings in domestic architecture which are in some way the consequence of the use of reinforced concrete? [. . .] M. Arnaud has doubtless not dared to risk such an undertaking [. . .]. How much more edifying his façade would have been had he just made the effort to adorn it in its own way, extracting from the study of his material the elements of an entirely personal decoration of his own design.[25]

Once again, the stage had been reached in the exploitation of a new material at which concerns for authenticity demanded its undisguised use and the acceptance of the aesthetic principles that emerged from such use.

Reinforced concrete achieved its aesthetic maturity in the work of Auguste Perret.[26] One of his earlier works is the apartment block at 25b, rue Franklin, Paris, built in 1903 (fig. 5). The concrete skeleton of this building has the advantage of removing from the plan of the apartments any

25. Quoted from Collins (1959), p. 70.
26. On Perret see Collins (1959), pp. 153–287, and Banham (1960), pp. 38–43.

Induction and Revolution in the Applied Arts

Figure 5. Auguste Perret, apartment block, 25b, rue Franklin, Paris (1903). The British Architectural Library, RIBA, London.

load-bearing walls. But the façade, clad in ceramic tiles, appears as yet unwilling to acknowledge the material that dictates the building's form. Soon, however, Perret came to reject such decoration, and allowed the appearance of his buildings to reflect their concrete frame. For example, the Admiralty Research Laboratories, on boulevard Victor, Paris (1928), are simple rectangular buildings with blank walls in which the structural elements are displayed openly. Even more influential architecturally were the several churches that Perret designed. Typical of these is the church of Notre Dame at Le Raincy, near Paris (1922), in which visible concrete columns and vaulted slabs frame large expanses of glass (fig. 6). Even at this late date, some architectural critics objected to Perret's design, maintaining that concrete is insufficiently noble for churches and should have been covered by a decorative cladding. Nonetheless, reinforced concrete had by that time generally attained architectural acceptance in virtue of its aesthetic merits as well as of its utilitarian benefits: from then on, one can speak of an aesthetic of reinforced concrete.[27]

Other building materials, such as aluminum and plate glass, passed through the same stages as cast iron and reinforced concrete before gaining aesthetic acceptance. Virtually every important building material now in use demonstrated its utilitarian benefits on the fringes of architectural activity before the forms most suited to its exploitation came to be accepted by architectural taste.

4. Materials and Forms in Industrial Design

The results of our two case studies in the history of architecture hold also in industrial design.[28] Here too, a new material is generally used first in designs that have been devised to exploit longer-established materials. Only gradually do manufacturers give their products the forms that are best suited to the new material; and then only gradually do these forms win acceptance among customers. Eventually, of course, the customers may come to expect such objects to have no form other than the one introduced by the new material.

An example of this development is provided by steel, which became available for household goods in the second half of the nineteenth century. Early designs for steel furniture tended to imitate the forms made familiar by the traditional material, wood: only in the 1920s did furniture

27. A discussion of some aesthetic principles underlying the use of reinforced concrete in architecture is Michelis (1963).

28. On materials and form in industrial design see Heskett (1980) and Sparke (1986a), especially pp. 37–55, 124–139.

Figure 6. Auguste Perret, Notre Dame, Le Raincy (1922). The British Architectural Library, RIBA, London.

design begin to explore the new forms that the material offered.[29] Similarly, when plastics became available for use in consumer goods in the 1920s, they were initially seen as economical substitutes for traditional materials, with no effect on design. For example, the early plastic buttons, buckles, and combs were replicas of similar articles in wood, horn, and ivory. The new designs that were made possible by plastics emerged only in the 1930s in such objects as portable radios: these new forms were gradually attributed aesthetic value of their own.

When in the early days of the exploitation of a new material manufacturers give it traditional forms, neglecting to pursue its distinguishing aesthetic possibilities, they may please aesthetically conservative consumers but do not win praise from those who hold authenticity in high regard. Nikolaus Pevsner lists some of his dislikes: "In a cardboard travelling-case made to imitate alligator skin, in a bakelite hair-brush made to imitate enamel—there is something dishonest. A pressed-glass bowl trying to look like crystal, a machine-made coal-scuttle trying to look hand-beaten, machine-made mouldings on furniture, a tricky device to make an electric fire look like a flickering coke fire, a metal bedstead masquerading as wood—all that is immoral."[30] The form imposed on each of these articles appears to disavow its new material, and this may be seen as a form of aesthetic betrayal.

5. THE INDUCTION TO STYLES

As these case studies show, the process by which aesthetic canons in the applied arts respond to improvements in technical capability closely resembles the inductive mechanism that, I have suggested, gives rise to aesthetic canons for theory choice in science. Consider the similarities between the two processes.

Aesthetic canons in architecture respond to two factors: the aesthetic features of past buildings and their perceived utilitarian worth. The utilitarian worth of a building, and of others aesthetically similar to it, determines the value that the aesthetic canon attributes to the features that the building exhibits; in turn, the canon is used both to guide and to assess the design of subsequent buildings. A well-entrenched aesthetic canon will ensure that the community designs and esteems buildings that are aesthetically orthodox. Orthodoxy of design will continue to be valued when a new material becomes available. When a building is erected that

29. On furniture design in the machine age, see Sparke (1986b), pp. 26–51.
30. Pevsner (1937), p. 11.

shows aesthetic features made possible by and appropriate to a new material, it is at first disliked, on the strength of the established aesthetic canon. Only when this building, or others aesthetically similar to it, shows sufficient utilitarian worth does the weighting of its aesthetic features within the canon increase appreciably. This occurs especially if the new building meets needs that cannot be satisfied by buildings of more orthodox aesthetic form. This change in weightings allows buildings showing the new features to win acceptance on aesthetic as well as utilitarian grounds. The revision of the canon ensures that credit will more likely be extended to future buildings that embody the new aesthetic features, enabling architects further to exploit the new material.

Similarly, on the model that I have been developing, aesthetic canons in science pay regard to two factors: the empirical success of theories embraced by the community and their aesthetic properties. The empirical success of a theory contributes to determine the weighting of that theory's aesthetic properties within the community's aesthetic canon; in turn, this canon is used to evaluate subsequent theories. An entrenched aesthetic canon will cause the community to produce and esteem aesthetically orthodox theories. Sometimes a theory emerges which, perhaps in consequence of new assumptions or techniques, shows unprecedented aesthetic properties. Such a theory is likely at first to be resisted in virtue of the established aesthetic canon. Only when this theory, or others similar to it, has shown sufficient empirical success does the weighting of its aesthetic properties rise substantially. This allows the new theory to win acceptance on aesthetic as well as empirical criteria. The revision of the canon ensures that subsequent theories showing the new aesthetic properties are more likely to be accepted, enabling the community to pursue further the assumptions or techniques that constitute the new style of theorizing.

In both domains, therefore, the demonstrated practical worth of a work—empirical success in the case of scientific theories, utility in the case of buildings—is capable of reshaping the aesthetic canons on which subsequent work is evaluated and by which the line of progress of the discipline is partly determined. Consequently, aesthetic canons show the same distinctive life cycle in science and the applied arts: an aesthetic canon arises in response to the empirical success of a sequence of aesthetically innovative contributions; it initially meets with opposition from those who maintain allegiance to the previously established canon; it enjoys a period of influence over the community's preferences; and it eventually comes to hamper further empirical progress, by ruling subsequent aesthetically innovative work as unacceptable.

A scientific theory that stands in a particular relation to an aesthetic canon at a certain moment in its life cycle has its counterpart in a work of applied art that stands in a similar relation to the aesthetic canons in that art form. For example, Copernican theory stands to the aesthetic canon that prevailed in mid-sixteenth-century mathematical astronomy as a mid-eighteenth-century masonry building stands to the aesthetic canon that prevailed in architecture at that time: both entities accord fully with the well-established aesthetic canons in their field. The early version of quantum theory, to which Planck and Einstein could still adhere, occupies within the twentieth-century revolution in physics a place analogous to that of Labrouste's Bibliothèque Sainte-Geneviève in the rise of iron designs: each contains elements of profound innovation but retains enough of the appearance of the long-established style to appeal to aesthetic conservatives.

The similarities between the processes by which aesthetic canons develop and are updated in science and applied art should dissipate any skepticism about whether one can properly regard as aesthetic an evaluative canon formed by an inductive response to empirical performance. We now see that aesthetic canons in the applied arts also respond to empirical performance. Moreover, the manner of this response in the applied arts closely resembles the inductive mechanism operating in the sciences. So the claim that scientists' aesthetic canons are shaped partly by their theories' empirical success, far from conflicting with the usual understanding of aesthetic preferences, accords with the situation in the applied arts.

We saw in the previous chapter that the aesthetic properties of empirically successful theories may change faster than the aesthetic canon can develop: these are the circumstances in which a revolutionary crisis may arise. The same phenomenon can occur in the applied arts. When the rate of progress of technical capabilities in the applied arts is high, aesthetic canons may fail to renew themselves quickly enough to ensure that they will accommodate the new capabilities, giving them a form that the canon regards as seemly. In the following passage, the architect Maxwell Fry discusses the implications for architectural design of the high rate of technical progress maintained during the nineteenth century: "The rapidity of this change cut the ground away from under the architect's feet. [. . .] If the structural developments which have led to our present technical skill were to continue at the same pace into this century, at a pace, that is, exceeding our capacity as artists to assimilate them, then our hopes of establishing a workable architecture would be slight."[31] The possible inability of architectural canons to keep pace with technical de-

31. Fry (1944), p. 122.

Induction and Revolution in the Applied Arts

velopments is analogous to the lag of aesthetic canons in a science in a revolutionary crisis.

Each of the models that we have discussed in the past two chapters—my model of the evolution of aesthetic canons in science, the Marxist model of history, and the above account of the development of aesthetic canons in the applied arts—may be read either as a deterministic, single-factor account of its subject matter or as a component of an adaptable, multifactor account. In each instance, the deterministic, single-factor reading is unconvincing: in all the fields that we have considered, individual creativity constitutes a source of innovation that cannot be predicted or explained by the progress of technical capabilities. No present-day advocate of the Marxist model of history would claim that a particular set of productive capabilities leads inescapably to a particular mode of organization: the organization of a society is determined partly by choices made by its members. The development of cast iron in architecture gave rise to the sensuality of Art Nouveau as well as to the functionalism of the Eiffel Tower. In industrial design, concerns for authenticity coexist with a delight in the *faux*, expressed in paste jewelery and fake fur. Similarly, the model of the evolution of aesthetic canons in science that I give should not be interpreted deterministically, as though there existed iron laws of scientific development. On the contrary, in their choice of theories scientists may exercise aesthetic preferences that are not imposed by the aesthetic induction.

All the models that we have discussed should thus be interpreted primarily as identifying a trend in the developments that they describe. This trend coexists and combines with other causal factors to yield the richness of the phenomena that we observe—the evolution of societies, the history of applied art, and the progress of science.

Circles and Ellipses
in Astronomy

1. TESTING THE MODEL AGAINST HISTORY

The model of scientific practice that I have developed can now be tested against historical evidence. Philosophical models of science need not agree in every detail with the interpretations that historians have given of episodes of the history of science: indeed, they may provide reasons for revising these interpretations. But they should be broadly consistent with, and if possible account for, documentary evidence of historical occurrences.

Not all historical episodes will constitute equally significant tests of a model of scientific practice: the more significant tests are offered by the episodes about which the model's claims are most distinctive. In the case of the model developed here, these episodes are scientific revolutions. As we saw in Chapter 8, in a nonrevolutionary period in the history of a science, a community's aesthetic canon for theory choice accords with its empirical criteria for theory choice. A scientific revolution occurs when theory choices performed on empirical criteria depart from those performed on the aesthetic canon. At such times, the aesthetic canon escapes from the shadow of empirical criteria: it exercises a conservative function, advocating the retention of theories showing familiar aesthetic properties and the rejection of their new, aesthetically innovative competitors.

In this chapter and the next, I test my model of scientific practice against two pairs of historical episodes: the rise of Copernicus's theory and Kepler's theory in mathematical astronomy and the rise of relativity theory and quantum theory in physics.[1] Each of these four episodes is commonly considered revolutionary. We will discover, however, that in the case of the first episode of each pair, the theory that arose showed agreement with an established aesthetic canon. Copernicanism fulfilled a long-standing requirement placed on theories in mathematical astronomy, which may be described in terms of simplicity, symmetry, or metaphysical allegiance: the requirement that the motion of celestial bodies should be interpreted as uniform motion along circles or combinations of circles. Relativity theory, likewise, satisfied requirements that had become established in physics in the nineteenth century, that theories should be deterministic and show particular symmetries. Because these theories satisfied such requirements, neither should be considered revolutionary. On the contrary, I would argue, Copernicanism is most appropriately seen as the culmination of Ptolemaic-style astronomy, and relativity theory should be seen as the culmination of classical physics.

I shall contrast each of these nonrevolutionary episodes with the other episode of its pair, which constituted a genuine aesthetic rupture and hence, on my view, a revolution. By describing planetary orbits as ellipses, Kepler disavowed the commitment to uniform circular motions to which mathematical astronomy had held; similarly, the developers of quantum theory abandoned the commitment that Planck and Einstein had to determinism and that of Schrödinger to visualization. I will show that, in these episodes, the scientists who resisted the revolution regarded the innovative theory of their time as aesthetically displeasing: followers of Ptolemy and Copernicus abhorred Kepler's ellipses as imperfect and inappropriate to celestial motions, while Planck and Einstein considered quantum theory aesthetically repugnant in consequence of its indeterminism. This is evidence that participants in a scientific revolution experience it as a rupture with established aesthetic canons.

2. Did Copernicus's Theory Constitute an Empirical Advance?

Most historians of science presume that mathematical astronomy underwent a revolution sometime between 1500, when Western astronomy was still dominated by the Ptolemaic theory of the heavens, and 1650, when Newton commenced his studies in mathematics. Which of the many innovations in mathematical astronomy during this time span deserves to be regarded as a revolution? The answer customarily given by historians since the mid–eighteenth century is that mathematical astronomy under-

1. Another version of this chapter is contained in McAllister (1996).

went a revolution at the hands of Copernicus, consisting in the transition from geocentrism to heliocentrism.[2] I dispute this claim.

I emphasize at once that we are here considering whether Copernicus's theory accomplished a revolution in mathematical astronomy, not in philosophy. Whether Copernicanism was philosophically revolutionary is a much debated question. Some writers have portrayed Copernicus as displacing humankind from the center of the universe to a peripheral position in a vast cosmos, and they have argued that this amounts to a revolution in philosophical anthropology; others, such as Arthur O. Lovejoy, have argued that Copernicanism had much less philosophical impact than this.[3] But these arguments have no bearing on the effect of Copernicus's theory in mathematical astronomy.

The Aristotelian classification of the sciences, which was endorsed up to the time of Copernicus, regarded mathematical astronomy as separate from physical cosmology. The former discipline was dedicated to developing mathematical models to predict the positions of celestial bodies; the latter was the branch of natural philosophy devoted to ascertaining the nature and causes of celestial motions. In consequence of this demarcation, the task of mathematical astronomers was not primarily to provide accounts of celestial phenomena that were literally true.[4] Copernicus's theory, like Ptolemy's, was put forward as a contribution to mathematical astronomy; thus, whether it constituted a revolution in its discipline must be gauged from its effects in mathematical astronomy, not its repercussions in physical cosmology.

To decide whether a particular scientific theory constituted a revolution in its discipline, we must ascertain the grounds on which the theory attracted adherents and opponents. A revolutionary theory attracts support on the strength of its empirical performance and opposition in virtue of its aesthetic properties; a theory that falls short of revolutionary attracts support on the strength of its aesthetic properties, whatever effect its empirical performance has on its reception.

According to some accounts, mid-sixteenth-century mathematical astronomers were prompted to transfer their allegiance from Ptolemy's theory to Copernicus's on empirical grounds, such as predictive accuracy and degree of simplicity. Let us begin by investigating the plausibility of this view. There are two categories of predictions on which Copernicus's and Ptolemy's theories could be compared: quantitative predictions of

2. For a history of the early historiography of the "Copernican revolution," see I. B. Cohen (1985), pp. 498–499.

3. Lovejoy (1936), pp. 99–108.

4. On the distinction between mathematical astronomy and physical cosmology, see Hanson (1961), pp. 170–172, and Jardine (1982), pp. 183–189.

Circles and Ellipses in Astronomy

the positions of celestial bodies, and qualitative predictions of the appearance of the night sky.

There is little evidence either that mathematical astronomers of Copernicus's time were dissatisfied with the accuracy of the quantitative predictions of Ptolemaic theory or that they considered Copernicus's theory more accurate. Copernicus pronounces himself content with the predictive accuracy of Ptolemaic theory at the opening of both the *Commentariolus*, a treatise which he composed probably between 1510 and 1514, and *De revolutionibus orbium coelestium*, the work of 1543 for which he is most famous.[5] The predictions of Ptolemaic and Copernican theory have been compared by several present-day scholars, who report the latter to be no more accurate than the former.[6] In fact, choosing between Ptolemaic and Copernican theory on grounds of predictive accuracy would have required astronomical data more precise than were available for decades after the publication of *De revolutionibus*. Thus, even if Copernicus's theory had been more accurate than Ptolemaic theory, this would not have been apparent to astronomers of the time. In sum, the claim advanced by some old historiography—that by the mid–sixteenth century Ptolemaic theory had led astronomy into an empirical crisis, which was resolved by Copernicus—is untenable.[7]

Copernicus's theory also failed to demonstrate superiority over Ptolemaic theory in qualitative predictions about the appearance of the heavens. For instance, many of his contemporaries reasoned that, if the earth were in motion, observations would show some stars appearing to oscillate with respect to others, by the effect of annual parallax. The fact that no such oscillations were observed weighed against Copernicus's theory. A second influential observation was that the brightness of Venus is approximately constant. This fact is at odds with Ptolemaic theory, which asserts that the distance of Venus from the earth fluctuates greatly, suggesting that we should observe corresponding fluctuations in brightness. However, Copernicus's theory could draw no advantage from this: for different reasons, it predicted similarly great fluctuations in the distance of Venus from the earth, and it offered no separate explanation of the constancy of the brightness.[8] The present-day explanation is that the

5. Swerdlow (1973), p. 434; Copernicus (1543), p. 4. I quote both the relevant passages later in this chapter.

6. The predictive accuracy of Ptolemaic and Copernican theory has been compared, for instance, by D. J. de S. Price (1959), pp. 209–212, Gingerich (1975), pp. 85–86, and I. B. Cohen (1985), pp. 117–119.

7. Gingerich (1975).

8. On the difficulty of Copernicus's theory in accounting for the brightness of Venus, see D. J. de S. Price (1959), pp. 212–214.

apparent luminosity of Venus depends both on its distance from the earth and on its phase, which happen to compensate for each other almost exactly. But the fact that Venus shows phases was first discovered in 1610 by Galileo Galilei. Even this news failed to establish the superiority of Copernican theory, since some versions of Aristotelian cosmology also suggested that Venus showed phases.[9]

On grounds such as these, Robert Palter concludes that Copernicus's theory was not perceptibly superior to Ptolemy's in predictive accuracy. "In order to square this fact with the putative reality of a 'Copernican revolution'," according to Palter, "one is constrained to fall back on the criterion of simplicity."[10] If Copernicus's theory was simpler than Ptolemy's, it could be regarded as empirically superior, even though it offered no greater predictive accuracy.

That Copernicus's theory attracted support mainly in virtue of the degree of its simplicity has been suggested by many historians and philosophers of science.[11] Most present-day estimates of the relative degrees of simplicity of Ptolemaic and Copernican theory are based on a tally of the circles to which the theories appeal: it is frequently claimed that whereas Ptolemy used eighty-odd circles, Copernicus's theory requires only thirty or so.[12] However, such a tally does not reflect the simplicity that mathematical astronomers of the sixteenth century regarded as significant. Their typical problem was to calculate the apparent position of a planet from the earth. No problem of this sort required the use of all the eighty-odd circles of Ptolemaic theory: it needed no more than the six or so circles governing the motion of the planet to which the problem refers. On Copernicus's theory, by contrast, both the earth and the other planet are moving, so the problem involves the circles governing the motions of both bodies. In this sense, as a set of solutions to individual problems, Ptolemaic theory is simpler and more convenient—if somewhat less systematic—than that of Copernicus.[13]

In fact, Copernicus's theory was not generally regarded as simpler

9. On the role of discussions about the phases of Venus in the controversy over Copernicanism, see Ariew (1987).

10. Palter (1970), pp. 114–115. Palter also undermines the suggestion that Copernicus's theory was superior to Ptolemy's in physical plausibility.

11. For instance, Reichenbach (1927), p. 18, writes: "Copernicus [. . .] was able, in fact, to cite as a distinct advantage only the greater simplicity of his system."

12. This is the view of, for example, Kordig (1971), p. 109, who states that Copernicus simplified Ptolemaic astronomy by reducing the number of epicycles "from 84 to about 30." For further details and examples of the tally of circles see Palter (1970), pp. 94, 113–114, and I. B. Cohen (1985), p. 119.

13. For further discussion of the degrees to which Ptolemy's and Copernicus's theories were simple and systematic, see Hanson (1961), pp. 175–177.

Circles and Ellipses in Astronomy

than Ptolemaic theory at the time of its formulation.[14] There is evidence that Copernicus, too, eventually realized that he could claim on behalf of his theory no greater simplicity than that of Ptolemaic theory. Although in the *Commentariolus* Copernicus had suggested that his theory was simpler, in the more methodical *De revolutionibus* he claimed instead that it had superior internal harmony.[15]

Predictive accuracy and degree of simplicity thus appear not to be grounds on which Copernicus's theory could win adherents from Ptolemaic theory. To discover on what grounds Copernicus's theory proved attractive, we must bring to light the problem in mathematical astronomy that it was designed to solve.

3. COPERNICUS'S RETURN TO ARISTOTELIAN PRINCIPLES

In the fourth century B.C., Aristotle had enunciated three principles in physical cosmology. The first was the principle of centrality and immobility of the earth. The second was the principle of bipartition of the universe, which stated that the sublunar region, containing the earth and its atmosphere, differs in physical nature from the supralunar region, consisting of the remainder of the universe. The third was the principle of circularity and uniformity of celestial motions, which stated that celestial bodies move with uniform linear velocities along paths that are circles or compounds of circles. The latter two principles, in particular, were deeply embedded in Aristotelian natural philosophy. This held that objects of the sublunar region, composed of the four elements traditionally cited in ancient cosmologies, are subject to violent or forced motions that displace them from their natural locations. By contrast, celestial bodies are composed of a fifth element or quintessence, ether, which confers perfection and ensures that they move only with motions natural to them, namely uniform circular motions.[16]

Since Aristotle never formulated a theory in mathematical astronomy, his followers were anxious to devise one that accorded as fully as possible with his cosmological principles. Such astronomers as Apollonius of Perga (third century B.C.) and Hipparchus (second century B.C.) de-

14. Further evidence that, at the time of its formulation, Copernicus's theory was not considered simpler than Ptolemaic theory is given by Neugebauer (1968).

15. On the claim to simplicity in the *Commentariolus*, see Swerdlow (1973), pp. 434–436. The replacement of this claim by the claim to internal harmony in *De revolutionibus* is noted by Pera (1981), pp. 157–159.

16. For further discussion of the doctrine of uniform circular motions in Aristotle's cosmology, see Randall (1960), pp. 153–162.

scribed the motions of celestial bodies by appeal to systems of circles centered at least roughly on the earth. Their theories thus satisfied the principle of the earth's centrality and immobility and that of circularity and uniformity of celestial motions, and they did not conflict with Aristotle's principle of bipartition of the universe. The chief difficulty encountered by astronomers in this tradition was in accounting satisfactorily for observational data. Several times it occurred that a theory was found incapable of accounting for the data to acceptable accuracy, and a more intricate arrangement of circles had to be devised to improve the fit. Eventually, around A.D. 150, Claudius Ptolemy in the *Almagest* concluded that satisfactory accord with the data required a new geometrical device: the equant point.

Consider all cases of a body moving along a circle at such a rate that its angular velocity about some point is uniform, and call this point, as did Ptolemy, the equant point. In the case in which the equant point coincides with the center of the circle, the body has also uniform linear velocity along its circular path. In formulating a model of celestial motions, one may stipulate that the equant point governing a body's motion should coincide with the center of the circle along which the body travels: this is what, in effect, Ptolemy's predecessors had done in stating that celestial bodies travel with uniform linear velocity. By contrast, Ptolemy allowed himself the freedom of locating the equant point so as to optimize the model's fit with the data: in this case, the equant point will coincide with the center of the circle only rarely.

Partly thanks to this degree of freedom, Ptolemy's theory was much better than its predecessors at accounting for astronomical data; on the strength of this, it dominated Western mathematical astronomy until the sixteenth century. Use of the equant point amounted to relaxing somewhat the commitment to the principle of circularity and uniformity of celestial motions, however, since celestial bodies were no longer represented as moving in their orbits with uniform linear velocities. The equant point was criticized for this reason by natural philosophers from Proclus in the fifth century to Girolamo Fracastoro in the sixteenth.[17] The fact that the best-performing theory in mathematical astronomy available, the Ptolemaic theory, was not fully consistent with the fundamental theory in physical cosmology, based on Aristotle's three principles, was widely deplored in medieval natural philosophy.[18]

17. On the misgivings of Proclus and Fracastoro about the equant point, see Hallyn (1987), p. 120 and notes.
18. On the tension between Ptolemaic mathematical astronomy and Aristotelian physical cosmology in the Middle Ages, see Grant (1978), pp. 280–284.

Circles and Ellipses in Astronomy

Copernicus shared the dissatisfaction with Ptolemy's recourse to the equant point. He considered that theories in mathematical astronomy should accord fully with the principle of circularity and uniformity of celestial motions. This conviction is evident in one of the chapter titles of *De revolutionibus*: "The motion of the heavenly bodies is uniform, eternal, and circular or compounded of circular motions."[19] The equant point violated this principle, and Copernicus wished to rid astronomical theory of it. This intention is visible as clearly in his arguments against Ptolemy as in his own theorizing. First, he criticized Ptolemaic theory in both the *Commentariolus* and *De revolutionibus* not as a supporter of heliocentrism criticizing geocentrism but on the grounds that Ptolemy had adhered insufficiently strictly to the principle of circularity and uniformity of celestial motions. Second, he constructed a theory which, by avoiding use of equant points, more fully satisfied the principle of circularity and uniformity of celestial motions, and conformed to the principle of bipartition of the universe as well. He retraced his reasoning at the opening of the *Commentariolus*:

> The theories concerning these matters that have been put forth far and wide by Ptolemy and most others, although they correspond numerically [with the apparent motions], also seemed quite doubtful, for these theories were inadequate unless they also envisioned certain equant circles, on account of which it appeared that the planet never moves with uniform velocity either in its deferent sphere or with respect to its proper center. Therefore a theory of this kind seemed neither perfect enough nor sufficiently in accordance with reason.
>
> Therefore, when I noticed these [difficulties], I often pondered whether perhaps a more reasonable model composed of circles could be found from which every apparent irregularity would follow while everything in itself moved uniformly, just as the principle of perfect motion requires.[20]

In other words, Copernicus sought to formulate a theory in mathematical astronomy that accorded with Aristotelian physical cosmology more closely than Ptolemy's did.

Copernicus's theory, it is true, involved violating the principle of the earth's centrality and immobility; and this departure from orthodoxy attracted its share of criticism from Aristotelian natural philosophers.[21]

19. Copernicus (1543), p. 10. For further discussion of the status of the circle in sixteenth-century astronomy, see Brackenridge (1982), pp. 118–121.
20. Swerdlow (1973), pp. 434–435; interpolations by Swerdlow.
21. On Aristotelian objections to Copernicus's claim of the motion of the earth, see Grant (1984).

However, Copernicus's attribution of motion to the earth met less resistance than we might expect, for two reasons. The first was that the demarcation between mathematical astronomy and physical cosmology allowed Copernicus's readers, if they so chose, to entertain his theory as a mathematical model to predict the positions of celestial bodies, without acquiring a commitment to the claim that the earth is truly in motion. This was the stance urged on readers by the unsigned preface that Andreas Osiander added to *De revolutionibus*.[22]

The second fact that diminished the resistance to Copernicus's suggestion was that the principle of the earth's centrality and immobility had always been the most widely disputed of the three principles of Aristotelian cosmology. Pythagorean astronomers, such as Aristarchus of Samos in the third century B.C., had rejected it for heliocentrism.[23] Forms of sun worship, which retained their popularity into the Renaissance, also supported heliocentrism.[24] Another astronomical theory, first propounded by Heraclides of Pontus in the fourth century B.C., conflicted with the spirit of Aristotle's principle by asserting that the sun, moon, and outer planets orbit the earth, but that Mercury and Venus orbit the sun, thus admitting a second center of rotation in the universe. This theory had been widely endorsed by educated people throughout the Middle Ages. Many of those who had denied that the earth was at the center of the universe also attributed motions to the earth, and maintained that the hypothesis that the earth moves does not conflict with experience as severely as it superficially appears to. Copernicus thus found numerous precedents for his suggestion: in *De revolutionibus* he cites Pythagoreans as holding to heliocentrism, and Hicetas of Syracuse and other ancient astronomers as attributing motion to the earth.[25] He was thereby able to portray himself as merely defending views that had long been in circulation.

4. THE AESTHETIC PREFERENCE FOR COPERNICUS'S THEORY

As we have seen, by attributing uniform circular motions to celestial bodies, Copernicus's theory fulfilled a requirement of Aristotelian physical cosmology. How did mid-sixteenth-century astronomers perceive this

22. Copernicus (1543), p. xvi.
23. On Pythagorean heliocentrism, see Heninger (1974), pp. 127–128.
24. On forms of sun worship in the Renaissance, see Hallyn (1987), pp. 127–147.
25. Copernicus (1543), pp. 4–5 and p. 12. For discussion of Copernicus's appeals to ancient astronomers, see Hallyn (1987), pp. 59–62; for further references on Copernicus's Pythagoreanism, see Hallyn (1987), pp. 304–305 n. 25.

property of the theory? I suggest that they perceived it as giving aptness to the theory: they saw it as apt that celestial bodies should be described as having uniform circular motions.

Copernicus, too, saw his theory as apt. In *De revolutionibus* he claims as the chief merit of his theory an internal harmony greater than that of Ptolemaic theory:

> Those who devised the eccentrics seem thereby in large measure to have solved the problem of the apparent motions with appropriate calculations. But meanwhile they introduced a good many ideas which apparently contradict the first principles of uniform motion. Nor could they elicit or deduce from the eccentrics the principal consideration, that is, the structure of the universe and the true symmetry of its parts. On the contrary, their experience was just like some one taking from various places hands, feet, a head, and other pieces, very well depicted, it may be, but not for the representation of a single person; since these fragments would not belong to one another at all, a monster rather than a man would be put together from them.[26]

Copernicus expected that the aptness of his theory would prompt astronomers to transfer their allegiance from Ptolemaic theory to his own, even though it did not demonstrate any clear empirical superiority. This expectation proved largely correct. There is good evidence that many late-sixteenth-century mathematical astronomers found the aptness of Copernican theory attractive enough to outweigh any reservations that they may have had against the theory on other grounds. Examples of this attitude are offered by Erasmus Reinhold, Georg Joachim Rheticus, and Tycho Brahe. Reinhold, professor of astronomy at Wittenberg and one of the leading mathematical astronomers of his time, shared Copernicus's view that theories in their discipline should attribute only uniform circular motions to celestial bodies, as we may surmise from the motto that he inscribed on the title page of his copy of *De revolutionibus*: "The axiom of astronomy: celestial motion is uniform and circular or composed of uniform and circular elements."[27] Reinhold commended Copernicus's theory for satisfying this requirement by abolishing the equant point; the fact that Copernicus placed the sun instead of the earth at the center

26. Copernicus (1543), p. 4; see also p. 22. For a more detailed discussion of the passage quoted here, see Westman (1990), pp. 179–182; on the body metaphor in Renaissance aesthetic thought, see Hallyn (1987), pp. 94–103; for further discussion of Copernicus's notion of harmony, see Rose (1975).

27. Quoted from Gingerich (1973), p. 58; Reinhold's attitude to Copernican theory is further documented ibid., pp. 55–59.

of the universe appears not to have troubled him appreciably. Rheticus, professor of mathematics at Wittenberg, spent some time studying with Copernicus before publishing an account of the latter's theory in 1540 under the title *Narratio prima*. Rheticus wrote:

> You see that here in the case of the moon we are liberated from an equant by the assumption of this theory, which, moreover, corresponds to experience and all the observations. My teacher dispenses with equants for the other planets as well, by assigning to each of the three superior planets only one epicycle and eccentric; each of these moves uniformly about its own center [. . .]. My teacher saw that only on this theory could all the circles in the universe be satisfactorily made to revolve uniformly and regularly about their own centers, and not about other centers—an essential property of circular motion.[28]

Tycho expressed similar feelings in a letter of 1587 to the astronomer Christoph Rothmann:

> Copernicus [. . .] had the most perfect understanding of the geometrical and arithmetical requisites for building up this discipline [of astronomy]. Nor was he in this respect inferior to Ptolemy; on the contrary, he surpassed him greatly in certain fields, particularly as far as the device of fitness and compendious harmony in hypotheses is concerned. And his apparently absurd opinion that the Earth revolves does not obstruct this estimate, because a circular motion designed to go on uniformly about another point than the very center of the circle, as actually found in the Ptolemaic hypotheses of all the planets except that of the Sun, offends against the very basic principles of our discipline in a far more absurd and intolerable way than does the attributing to the Earth one motion or another which, being a natural motion, turns out to be imperceptible. There does not at all arise from this assumption so many unsuitable consequences as most people think.[29]

As I suggested in Chapter 2, properties of a theory that are regarded as giving it aptness are best interpreted as aesthetic properties. If this is correct, the property of attributing uniform circular motions to celestial bodies is an aesthetic property, and astronomers who were attracted to

28. Rheticus (1540), pp. 135, 137.
29. Quoted from Moesgaard (1972), p. 38. Further praise of Copernicus by Tycho along these lines is reproduced in Hallyn (1987), p. 123.

Circles and Ellipses in Astronomy

Copernicus's theory on the strength of this property were swayed by aesthetic preferences.[30]

In fact, the conviction that celestial motions are circular was based partly on aesthetic considerations in ancient thought too.[31] The interpretation of Copernicus and his contemporaries that I am advancing thus portrays them as doing nothing stranger than reiterating long-standing aesthetic preferences in Western astronomy.

The sixteenth century's strong preference for theories that attribute uniform circular motions to celestial bodies is easily explained by the aesthetic induction. To Copernicus and his contemporaries, it would have seemed that Aristotelian natural philosophy had built up an impressive empirical track record. This appraisal was largely justified: contrary to the claims of propagandists for the new science, such as Descartes and Galileo, everyday experience accords well with Aristotelian theories in mechanics and biology. Through the operation of the aesthetic induction, the community came to attach great weight to the requirement that theories in the various sciences should show allegiance to the metaphysical claims of Aristotelianism. In mathematical astronomy, Copernicus's theory satisfied this requirement more fully than did Ptolemaic theory. This was sufficient for the community to regard Copernicus's theory as preferable to Ptolemy's, even though it demonstrated no clear empirical superiority.

According to the model of scientific revolution that I have presented, a theory that satisfies the aesthetic criteria to which the community holds at the time of its formulation is not a revolutionary theory. My model thus portrays Copernicus's theory as constituting less than a revolution in mathematical astronomy.[32] This conclusion accords with the view that most contemporary astronomers held of the theory: they regarded it either as an attempt to return mathematical astronomy to its ancient state or as a revival of Pythagoreanism.[33]

30. That Copernicus's theory gathered support primarily in virtue of its aesthetic properties is acknowledged by Neugebauer (1968), p. 103, Gingerich (1975), pp. 89–90, Hutchison (1987), pp. 109–136, and Westman (1990), pp. 171–172. Neyman (1974), p. 9, writes that "Copernicus introduced a completely novel yardstick for appraising a new theory: conformity with observations and intellectual elegance."

31. On aesthetic considerations in ancient astronomy, see Haas (1909), pp. 93–102.

32. Among historical accounts that argue that Copernicus's theory did not constitute a revolution in mathematical astronomy are those of Neugebauer (1952), p. 206, Hanson (1961), and I. B. Cohen (1985), pp. 123–125.

33. On the perception of Copernicus by his contemporaries as a restorer of astronomy, see Hallyn (1987), pp. 59, 302 n. 12; as a Pythagorean, see Heninger (1974), p. 130.

5. Kuhn's Account of the Acceptance of Copernicanism

Kuhn gives an account of the transition from Ptolemy's theory to Copernicus's that differs from mine. Although he raises the question "whether Copernicus is really the last of the ancient or the first of the modern astronomers," Kuhn endorses the view that Copernicus was responsible for a revolution in mathematical astronomy.[34] He therefore expects to find in the transition from Ptolemy's theory to Copernicus's the features that he attributes to revolutions. Kuhn's model of revolutions, as we have seen, suggests that a revolutionary theory does not easily persuade scientists on empirical grounds but typically attracts them on aesthetic criteria.

According to Kuhn, Copernicus's theory could not have won adherents from Ptolemaic theory on the grounds of either predictive accuracy or degree of simplicity: "Judged on purely practical grounds, Copernicus' new planetary system was a failure; it was neither more accurate nor significantly simpler than its Ptolemaic predecessors."[35] Rather, Kuhn believes that Copernican theory gained adherents on the strength of its aesthetic properties. According to him, the arguments advanced in *De revolutionibus* show that Copernicus himself was aware that he could attract astronomers to his theory most effectively by emphasizing its aesthetic properties:

Each argument cites an aspect of the appearances that can be explained by *either* the Ptolemaic *or* the Copernican system, and each then proceeds to point out how much more harmonious, coherent, and natural the Copernican explanation is. [. . .] Copernicus' arguments are not pragmatic. They appeal, if at all, not to the utilitarian sense of the practicing astronomer but to his aesthetic sense and to that alone. [. . .] The harmonies to which Copernicus' arguments pointed did not enable the astronomer to perform his job better. New harmonies did not increase accuracy or simplicity. Therefore they could and did appeal primarily to that limited and perhaps irrational subgroup of mathematical astronomers whose Neoplatonic ear for mathematical harmonies could not be obstructed by page after page of complex mathematics leading finally to numerical predictions scarcely better than those which they had known before.[36]

34. Kuhn (1957), p. 134; (1962), pp. 149–150. The quoted passage is from Kuhn (1957), p. 182.
35. Kuhn (1957), p. 171.
36. Ibid., p. 181; see also p. 172.

Circles and Ellipses in Astronomy

Kuhn concludes that Copernicus's theory established itself primarily in virtue of its aesthetic properties, despite its inability to demonstrate empirical superiority over Ptolemy's theory. He concludes therefore that the transition from Ptolemaic to Copernican theory exhibits the features that his model attributes to a scientific revolution.

As I have made clear, I share Kuhn's conviction that the acceptance of Copernicus's theory was determined primarily by aesthetic rather than empirical factors. But the historical evidence that we have reviewed suggests overwhelmingly that, on any reasonable construal of "scientific revolution," Copernicus's theory did not constitute a revolution in mathematical astronomy: on the contrary, it was both intended and received as a conservative contribution to the established paradigm in the discipline. In fact, Kuhn's own finding that astronomers switched from Ptolemy's to Copernicus's theory primarily under the impulse of aesthetic factors ought to persuade him that this episode constituted no revolution: Copernican theory was able to attract adherents through the appeal of its aesthetic properties precisely because of its conservatism, its fulfillment of aesthetic canons that had long shaped the preferences of mathematical astronomers. As Robert S. Westman puts it, far from being perceived as revolutionary, Copernicus's theory was respectfully welcomed into what Kuhn would term the normal science of sixteenth-century mathematical astronomy.[37]

In a test of Kuhn's model of revolutions and of mine, the transition from Ptolemy's to Copernicus's theory should therefore be cited not as an example of scientific revolution but as a nonrevolutionary episode. Interpreted thus, this episode casts some doubt on Kuhn's claim that revolutions are sparked by aesthetic factors that foster dissatisfaction with established theories: the course of mathematical astronomy from Ptolemy to Copernicus shows us that what Kuhn calls normal science is guided and maintained to a large extent by aesthetic preferences. But for a clearer test of Kuhn's claim that scientific revolutions are typically induced by aesthetic factors and inhibited by empirical factors, we require an episode that truly constituted a revolution. In the next section, I argue that the rise of Kepler's theory of planetary motion constituted a genuine revolution in mathematical astronomy. Having portrayed Copernicus as a revolutionary, Kuhn characterizes Kepler's theory as a "version of Copernicus' proposal."[38] In fact, acceptance of Kepler's theory required astronomers to abandon long-established commitments, as we shall now see.

37. Westman (1975), pp. 191–192.
38. Kuhn (1957), p. 219.

6. The Iconoclasm of Kepler's Ellipses

Kepler's *Astronomia nova* of 1609 set out his first two laws of planetary motion.[39] These were the fruit of his "war on Mars," the effort that he undertook between 1600 and 1605 to describe mathematically the motions of the sun's fourth planet. Kepler had at his disposal the planetary data gathered by his former employer, Tycho: these had an accuracy of around 1 percent, substantially higher than any previous body of astronomical data. The line of reasoning by which Kepler reached his first law—that every planet's orbit is an ellipse in which the sun is located at one focus—was strongly guided by empirical considerations. He proceeded, roughly speaking, by proposing various curves for the orbit of Mars and gauging the accord of each with Tycho's data.[40]

Kepler tested first the hypothesis that Mars moves in a circle. He found that its path would depart by as much as 8 percent from that observed by Tycho. In Kepler's view, this discrepancy was sufficiently large for a circular orbit to be ruled out.[41] Kepler now had the option of attempting to account for the data with greater precision by constructing combinations of circles, following the tradition stretching from Apollonius to Copernicus; but he chose a different option. The distribution of the discrepancies between the observed path of Mars and a circle suggested to Kepler in 1602 the figure that he should next consider: "The orbit is not a circle, but [passing from aphelion] enters in a little on either side [at quadratures] and goes out again to the breadth of the circle at perihelion, in a path of the sort called an oval." Kepler was unable to reconcile even this hypothesis with the data, however. He concluded in 1604 that the orbit is a curve contained between a circle and an oval, and in the same sentence he suggested which curve this is: "In the middle longitudes [. . .] the perfect circle prolongs [the true orbital path] by about 800 or 900 [parts in 152350, the mean radius of orbit] too much. My ovality curtails by about 400 too much. The truth is in the middle, though nearer to my ovality [. . .] just as though Mars's path were a perfect ellipse."[42] The first law of planetary motion that Kepler published in the *Astronomia nova* expresses this conclusion.

The role played by empirical factors in Kepler's own reasoning is evident: the theory that he published in 1609 was the empirically best-per-

39. Kepler (1609).
40. C. Wilson (1968) and Whiteside (1974) document the guiding role that empirical concerns played in Kepler's reasoning towards his first law.
41. On Kepler's conclusion that the orbit of Mars is not circular, see Whiteside (1974), pp. 6–7.
42. Quoted from Whiteside (1974), pp. 8, 11; interpolations by Whiteside.

Circles and Ellipses in Astronomy

forming of the candidates that he had examined. Now let us consider the roles played by empirical and other factors in the reception of Kepler's theory by mathematical astronomers.

Great metaphysical and aesthetic value was attributed to the circle in the early seventeenth century, just as it had been in the lifetime of Copernicus. The circle continued to be portrayed as a figure of the greatest significance in literary imagery, for instance.[43] In comparison the ellipse was perceived as aesthetically displeasing. Whereas today we usually describe the circle as a special case of an ellipse, in which the two axes have equal length, the sixteenth and early seventeenth centuries saw the ellipse as a distorted and imperfect circle.

This predilection for circles was shared by early-seventeenth-century astronomers, not excluding Kepler.[44] Many held that the circle was the only figure that could appropriately be attributed to celestial motions. For instance, Tycho had written the following to Kepler in 1599: "The orbits of the planets must be constructed exclusively from circular motions; otherwise they could not recur with a uniform and equal constancy, eternal duration would be impossible; moreover, the orbits would be less simple, would exhibit greater irregularities and would not be suitable for scientific treatment and practice."[45] In 1607 the astronomer David Fabricius, who had been kept informed by Kepler about his work, wrote to him in similar terms: "With your ellipse you abolish the circularity and uniformity of the motions, which appears to me the more absurd the more profoundly I think about it. [. . .] If you could only preserve the perfect circular orbit, and justify your elliptic orbit by another little epicycle, it would be much better."[46] Disapproval of Kepler's theory on such grounds persisted in the years following the publication of the *Astronomia nova*.

In contrast, it was difficult at first for astronomers to ascertain the empirical worth of Kepler's theory: they were much less familiar with the mathematical properties of the ellipse than with those of the circle, and could not easily derive from the theory predictions to test against astronomical observations. The theory's empirical worth became more obvious after 1627, when Kepler published the *Tabulae Rudolphinae*.[47] This is a compilation of tables and rules for predicting the positions of the

43. On circle imagery in seventeenth-century literature, see Nicolson (1950), pp. 47–80.

44. On the persistence of commitments to circular motions in Kepler's thought, see Brackenridge (1982).

45. Quoted from Mittelstrass (1972), p. 210.

46. Quoted from Koestler (1959), p. 347.

47. Kepler (1627).

moon and planets, based on Kepler's laws. In essence, it is a tabulation of the observational consequences of Kepler's theory, which thus opened itself to easy empirical test. Astronomers soon found the predictions set out in the *Tabulae Rudolphinae* to be in good agreement with observed planetary positions—even those of Mercury, the planet that had thus far proved most recalcitrant to astronomical models.[48]

Many contemporary astronomers were led by their experience with the *Tabulae Rudolphinae* to recognize that Kepler's theory had great empirical worth.[49] An example is Peter Crüger, professor of mathematics at Danzig, whose views are known from his correspondence with Philipp Müller, his counterpart at Leipzig. For some years after the publication of the *Astronomia nova*, Crüger had held an unfavorable opinion of Kepler's theory. He wrote, for instance, to Müller in 1624: "I do not subscribe to the hypotheses of Kepler. I trust that God will grant us some other way of arriving at the true theory of Mars." Once the *Tabulae Rudolphinae* had appeared, however, Crüger revised his opinion. In a letter to Müller of 1629, Crüger demonstrated the impact that the tables had made on his view of Kepler's theory:

> You hope that someone will give these tables [the astronomical tables of Longomontanus] a further polishing and you say that all astronomers would be grateful for this. But I should have thought that it would be a waste of time now that the Rudolphine Tables have been published, since all astronomers will undoubtedly use these. [. . .] I am wholly occupied with trying to understand the foundations upon which the Rudolphine rules and tables are based, and I am using for this purpose the Epitome of Astronomy previously published by Kepler as an introduction to the tables. This epitome which previously I had [. . .] so many times thrown aside, I now take up again and study [. . .]. I am no longer repelled by the elliptical form of the planetary orbits [. . .].[50]

This passage contains evidence about the grounds on which Kepler's theory attracted both support and opposition. I interpret its last sentence as indicating that, by 1629, Crüger had abandoned one of the criteria upon which he had previously given an unfavorable assessment of Kepler's theory: he now no longer opposes the theory on the grounds that it describes planetary orbits as elliptical. The reason for which Crüger feels

48. The esteem of seventeenth-century astronomers for the accuracy of the *Tabulae Rudolphinae* is documented in C. Wilson (1968), p. 24.

49. The effect of the *Tabulae Rudolphinae* on the reception accorded to Kepler's theory is described in J. L. Russell (1964), pp. 7–9.

50. The two passages of Crüger are quoted from J. L. Russell (1964), p. 8.

Circles and Ellipses in Astronomy

that he can no longer afford to reject Kepler's theory on these grounds is, as the first part of the passage makes clear, the high degree of empirical adequacy that the theory had manifested. It is to Crüger's credit that, after working with the *Tabulae Rudolphinae*, he was able to lay aside his initial extraempirical reservations against Kepler's theory and acknowledge that its empirical worth justified its adoption.

Not all astronomers shared Crüger's view that the theory's empirical performance outweighed what they regarded as its metaphysical and aesthetic shortcomings. One of the astronomers who most strongly supported the principle of circularity and uniformity of celestial motions in the seventeenth century was Galileo, whose readiness to reject entrenched beliefs did not extend to this issue. In the *Dialogue Concerning the Two Chief World Systems* of 1632 he wrote: "Only circular motion can naturally suit bodies which are integral parts of the universe as constituted in the best arrangement."[51] Accordingly, the choice that Galileo offers his readers is between two world systems that ascribe circular motions to celestial bodies, those of Ptolemy and Copernicus; he omits Kepler's suggestion that planets might move in some other curve, although he knew Kepler's work and had corresponded with him. Research by Alexandre Koyré and Erwin Panofsky has established that Galileo's conviction that the circle is the only appropriate path for celestial bodies is rooted in aesthetic preferences to which he clung.[52]

The model of scientific revolutions that I have presented predicts that the reception of a revolutionary theory will exhibit three phases. In the first, the theory incurs resistance because its novel aesthetic properties conflict with the community's canon. In the second phase, despite this resistance, the theory demonstrates empirical success greater than its competitors. In the third phase, the empirical success accumulated by the new theory is so substantial as to cause the greater part of the community—including members who were previously opposed to it on aesthetic grounds—to embrace it. To this end, members of the community will de-emphasize their established aesthetic preferences, so that these do not hamper acceptance of the new theory. The reception of Kepler's theory exhibits each of these phases. Most evident is the effect of the aesthetic properties of Kepler's theory: far from providing its appeal, they proved to be an obstacle to its acceptance, which had gradually to be overcome by demonstrations of the theory's empirical power.

51. Galilei (1632), p. 32.
52. Koyré (1939), p. 154; Koyré (1955); Panofsky (1954), pp. 20–28; Panofsky (1956), pp. 10–13. For further commentary on the role of aesthetic preferences in Galileo's thinking in astronomy, see Shea (1985).

Kepler's theory, unlike Copernicus's, constituted a revolution in its discipline. In fact, there is ample historical evidence, independent of any philosophical model of scientific practice, that Kepler's theory represents a much deeper innovation in mathematical astronomy than that of Copernicus. As Norwood R. Hanson puts it, "The line between Ptolemy and Copernicus is unbroken. The line between Copernicus and Newton is discontinuous, welded only by the mighty innovations of Kepler."[53]

53. Hanson (1961), p. 169.

Continuity and Revolution
in Twentieth-Century Physics

1. Two Flaws in Classical Physics

As the nineteenth century approached its close, it appeared to many that physical science had evolved into a structure of great beauty. The pillars of this structure were Newton's theory and its extensions in mechanics, Maxwell's theory in electrodynamics, and Boltzmann's theory in thermodynamics. The structure, unified by its commitments to visualization and determinism, seemed capable of accounting for all physical phenomena.

At the very end of the century, however, two flaws became apparent. Physicists described the flaws in various ways, depending on their viewpoints. Kelvin, in a lecture of 1900, identified them as follows: "The beauty and clearness of the dynamical theory, which asserts heat and light to be modes of motion, is at present obscured by two clouds. I. The first came into existence with the undulatory theory of light [. . .]; it involved the question, How could the earth move through an elastic solid, such as essentially is the luminiferous ether? II. The second is the Maxwell-Boltzmann doctrine regarding the partition of energy."[1] The first flaw consisted in an inability of physical theory to reconcile its accounts of motion and of electromagnetic phenomena; it manifested itself, as Kelvin notes, in the attribution of mutually inconsistent properties to the ether, the medium of propagation of electromagnetic radiation. The second flaw affected accounts of submicroscopic phenomena; it manifested itself, for example, in an inability to account for the way in which radiation energy is distributed over wavelengths.

1. W. Thomson (1901), p. 486.

In the years after 1900, physicists worked to repair these two flaws in physical science. The first was remedied by the development of a theory that shared the aesthetic properties of classical physics and recovered the beauty that nineteenth-century physicists had seen in it. Repairing the second flaw proved less straightforward: radical reform was required, in the course of which physical science lost some of its distinctive nineteenth-century features—to the great displeasure of some physicists. The remedy to the first flaw was provided by relativity theory; to the second by quantum theory.

Most accounts of twentieth-century physics portray both relativity and quantum theory as revolutionary. The model of scientific practice that I have been developing agrees that the rise of quantum theory constituted a revolution, since this theory failed to exhibit the aesthetic properties that the physics community customarily associated with empirical success. We shall see that the protagonists in the development of quantum theory offer good illustrations of the behavior of the progressive and conservative factions that form during revolutions. In contrast, relativity theory showed many notable aesthetic properties that nineteenth-century physicists had wished to see. On the strength of this fact, my model interprets the rise of relativity theory as nonrevolutionary. We begin by ascertaining the grounds on which relativity theory gained support.

2. Aesthetic Factors in the Appeal of Relativity Theory

Physicists at the end of the nineteenth century were alerted to the tension between their accounts of motion and of electromagnetic phenomena partly by empirical findings. In a series of experiments beginning in 1887, Albert A. Michelson and Edward W. Morley had used an interferometer to compare the speed of light in directions parallel and perpendicular to the motion of the earth in its orbit. Classical theory suggested that light beams sent in different directions with respect to the earth's motion travel at different speeds relative to the ether, but Michelson and Morley found no difference between the speeds of such beams. Since the hypothesis that the ether travels around with the earth in its orbit had to be rejected as implausible, it appeared that the speed of a light beam does not depend on the state of motion of the body emitting it. This result was at odds with Newtonian theory.

The tension between the classical accounts of motion and of electromagnetic phenomena did not become apparent only from empirical findings, however. Einstein's dissatisfaction with classical physical theory, for example, was not prompted primarily by empirical concerns.[2] Far from playing an important role in Einstein's thinking, Michelson and Morley's results came to his notice only some time after his paper putting forward the special theory of relativity had been published.[3] Indeed, that paper did not explicitly cite any of the recent experimental findings that cast doubt on classical physical theory.[4] Rather, Einstein's dissatisfaction with classical theory was motivated principally by factors that can be called metaphysical and aesthetic.

Einstein held strongly to a relationist view of space and motion that had originated in a criticism made of Newton by Leibniz and Mach. Newton had put forward an absolutist view of space and motion, according to which space is a physically concrete entity with respect to which objects are at rest or in motion absolutely. Leibniz and Mach argued that although we are able to ascertain empirically the velocities of objects relative to one another, there is no procedure to ascertain the absolute motion of an object. Therefore, if Newton's absolutism were correct, there would exist physical states of affairs that are distinct but empirically indistinguishable from one another, such as two systems of bodies that have the same motions relative to one another but different velocities in absolute space. In the light of his principle of the identity of indiscernibles, Leibniz felt that this consequence showed Newton's absolutism to be incorrect. Mach endorsed Leibniz's conclusion, arguing that the concepts of absolute space and motion were otiose, since Newton's own theories could be reformulated without them. On the Leibnizian-Machian view, physical states of affairs that are empirically indistinguishable should be regarded as physically equivalent and be given identical descriptions by physical theory.[5]

Einstein was motivated by Leibnizian-Machian relationism in his work on both the special and the general theory of relativity. He reacted in both cases to the fact that classical theory gave divergent descriptions of two physical systems in which all relative motions are identical and which he therefore regarded as physically equivalent. The divergence of these descriptions violated Leibnizian-Machian relationism. Each of his

2. For evidence of the limited role of empirical concerns in Einstein's thinking, see Swenson (1972), pp. 156–160, and Holton (1973), pp. 279–370.

3. For evidence that Einstein became aware of the Michelson-Morley results only after publication of his special relativity paper, see Swenson (1972), pp. 158–159, and Holton (1973), pp. 298–306.

4. On the role of experimental findings in Einstein's special relativity paper, see Holton (1973), pp. 306–309.

5. On Leibnizian-Machian relationism about space and motion and the influence that it had on Einstein, see Friedman (1983), pp. 3–17.

theories of relativity was an attempt of Einstein's to redescribe such physical systems by reference only to relative motions.

The constraints that Leibnizian-Machian relationism imposes on theorizing were expressed by Einstein mostly as symmetry requirements. These are evident in the 1905 paper in which Einstein put forward the special theory of relativity, "On the Electrodynamics of Moving Bodies." This paper opens with the following remark: "It is well known that Maxwell's electrodynamics—as usually understood at present—when applied to moving bodies, leads to asymmetries that do not seem to attach to the phenomena."[6] Einstein refers particularly to an asymmetry in the descriptions that classical theory gives of a system composed of a conductor and a magnet in motion relative to one another. Two cases may be distinguished: case A, in which the conductor is held fixed in some reference frame and the magnet is moved relative to it, and case B, in which the magnet is held fixed and the conductor is moved. In each case, there arises in the conductor an electric current of the same intensity. This has the consequence that an observer is unable to distinguish between the two cases by measuring the relative velocity of the bodies and the intensity of the current. But classical physical theory gives different explanations of the current in the two cases. In Einstein's words, "The observable phenomenon depends here only on the relative motion of conductor and magnet, while according to the customary conception the two cases, in which, respectively, either the one or the other of the two bodies is the one in motion, are to be strictly differentiated from each other."[7] Classical theory explains the current in case A by saying that the motion of the magnet produces an electric field that exerts a force on the electrons in the conductor; it explains the current in case B by saying that the electrons in the conductor experience a force as they move in the magnet's magnetic field. Thus, the account given of case B makes no appeal to the electric field that features in the account of case A.[8]

The special theory of relativity eliminates this asymmetry. One of its two postulates states that all inertial frames of reference are physically equivalent, from which it follows that an inertial frame can have no such

6. Einstein (1905), p. 140. On the concern for symmetry in Einstein's special relativity paper, see Holton (1973), pp. 380–385. As Holton points out (pp. 192–194), Einstein expresses displeasure at asymmetries in classical physical theory not only in his special relativity paper but also in his other two important papers of 1905. In contrast, Shelton (1988) argues that Holton overestimates the importance of symmetry considerations as a motivation of Einstein's work.

7. Einstein (1905), p. 140.

8. On the analysis given by classical theory of the magnet-conductor system, see Miller (1981), pp. 145–150.

Continuity and Revolution in Twentieth-Century Physics

property as an absolute state of motion. This postulate, which enjoins physical theories to refer to no motions other than the relative motions of bodies with respect to one another, thus disallows the account that classical theory gives of the magnet-conductor system. The second postulate of special relativity states that the speed of light is the same in all inertial frames. Einstein's paper demonstrates that these postulates are consistent with Maxwell's electrodynamics but inconsistent with Newtonian mechanics. The latter is replaced by a new or relativistic mechanics, which, conjoined with Maxwell's electrodynamics, gives an account of the magnet-conductor system that refers only to relative motions.

The fact that classical physical theory gives two different accounts of the magnet-conductor system cannot be considered an empirical failing: both accounts accord very well with observations. Einstein's dissatisfaction with classical physical theory, and his motivation in developing the special theory of relativity, arose primarily from a commitment to a metaphysical doctrine, relationism about space and motion. Einstein objected to the asymmetry in the accounts of the magnet-conductor system because he regarded it as a displeasing feature of the theory. This objection reveals what Abraham Pais sees as "relativity's aesthetic origins."[9]

Aesthetic factors are evident, too, in the reception accorded to the special theory of relativity. On the one hand, the theory was praised for accounting for the Michelson-Morley results; on the other hand, it was regarded as giving a more pleasing structure to physical science.[10]

The model of scientific practice that I have presented suggests that if a theory is able to win adherents on aesthetic as well as empirical grounds, we should not interpret its rise as a revolutionary event. To gain support on aesthetic grounds, a theory must accord with established aesthetic canons; it must therefore substantially share the aesthetic properties of previous theories. Einstein would have been the first to point out the continuity of his work with classical physics: "With respect to the theory of relativity it is not at all a question of a revolutionary act, but of a natural development of a line which can be pursued through centuries."[11] Several historians of twentieth-century physics agree that the special theory of relativity is most appropriately seen as the culmination of the program of classical physics. Holton, for instance, observes that "the so-called 'revolution' which Einstein is commonly said to have

9. Pais (1982), pp. 138–140. Swenson (1972), p. 157, concurs that the motivations that led Einstein to the special theory of relativity were largely aesthetic.

10. Wider issues in the reception of special relativity theory are examined in Glick (1987).

11. Quoted from Holton (1973), p. 197. For further evidence that Einstein regarded his work as maintaining continuity with classical physics, see Holton (1986), pp. 77–104.

introduced into the physics in 1905 turns out to be at bottom an effort to return to a classical purity. [. . .] Indeed, although it is usually stressed that Einstein challenged Newtonian physics in fundamental ways, the equally correct but neglected point is the number of methodological correspondences with earlier classics, for example, with the *Principia*."[12]

The special theory of relativity occupies in twentieth-century physics a place analogous to that of Copernicus's theory in sixteenth-century mathematical astronomy: each was intended and received as a contribution to an established style of theorizing that avoided the aesthetic imperfections of earlier contributions. Einstein's theory still fell short of his wishes in this regard, however. It conformed with Leibnizian-Machian relationism to the extent of declaring that all inertial frames of reference are physically equivalent, thereby dispensing with the concept of absolute velocity. It still distinguished, however, between frames that are inertial—i.e., unaccelerated—and those that are accelerated. In other words, it presumed that bodies have accelerations that are absolute rather than just relative to other bodies. Einstein strove to abolish the concept of absolute acceleration in the general theory of relativity.

The paper in which Einstein formulated the general theory opens in the familiar manner, by noting an asymmetry in the way existing theories described a particular physical system.[13] The system in question consists of two gaseous bodies in rotation relative to one another in an otherwise empty universe. Under particular conditions, it will be the case that body A is spherical while body B is oblate. What accounts for the difference in shape? Not the rotation of the bodies relative to one another, since that is a symmetric relation. Both Newtonian theory and the special theory of relativity would say that body B is oblate because it is in a state of absolute rotation, and body A is spherical because it is not in such a state. But this account, in virtue of referring to an absolute rotation, is at odds with Leibnizian-Machian relationism. The general theory of relativity was intended as an explanatory framework that would describe physical systems such as this by appeal only to relative motions.[14]

The general theory of relativity was mostly unsuccessful at providing such a framework; but the theory's failure to achieve its aims is of less interest to us than Einstein's motivation to develop it and the factors that determined its reception. Earlier (Chapters 1 and 6), I presented some testimony that Einstein was led to the general theory of relativity largely

12. Holton (1973), p. 195. For further support of the claim that the special theory of relativity maintained continuity with classical physics, see Hesse (1961), p. 226.

13. Einstein (1916), pp. 112–113.

14. The attempt and failure of the general theory of relativity to implement Leibnizian-Machian relationism are discussed by Friedman (1983), pp. 204–215.

Continuity and Revolution in Twentieth-Century Physics

by aesthetic considerations. Aesthetic criteria played an important part, alongside empirical criteria, in the theory's reception as well. In Peter G. Bergmann's opinion, "its eventual adoption, first by Einstein himself and later by the community of physicists, depended on the esthetic appeal of the finished theory and on its confirmation by experiment and observation."[15] Lorentz expressed the opinion that "Einstein's theory has the very highest degree of aesthetic merit: every lover of the beautiful must wish it to be true."[16] The conclusions that we drew for the special theory therefore hold also for the general theory: since it continued to show the symmetries and other properties valued by established aesthetic canons, it should be regarded as a contribution in the pre-existing style of theorizing rather than as a revolutionary innovation.

Aesthetic concerns were an equally strong factor in Einstein's work on a unified field theory, to which he devoted the latter part of his career. Pais has characterized Einstein's search for such a theory as follows: "Its purpose was neither to incorporate the unexplained nor to resolve any paradox. It was purely a quest for harmony."[17] Einstein never succeeded in formulating a unified field theory; but if he had, and if this theory had reflected the established aesthetic preferences of the community, it too would have counted on this model as nonrevolutionary.

3. Quantum Theory and the Loss of Visualization

While Einstein's theories of relativity won support partly on the strength of their aesthetic properties, quantum theory owed its early support almost entirely to its empirical properties: many physicists considered it aesthetically displeasing. Among the properties of quantum theory that were held against it were its lack of visualization and its metaphysical implications.[18]

Physical theory at the end of the nineteenth century offered systematic visualizations of submicroscopic phenomena. Electromagnetic radiation was visualized as a wave propagating in the ether, and various optical

15. Bergmann (1982), p. 30.
16. Lorentz (1920), p. 23. Among later physicists who have appreciated the aesthetic properties of the general theory of relativity are Chandrasekhar (1987), pp. 148–155, and Weinberg (1993), pp. 107–108.
17. Pais (1982), p. 23.
18. For further details of the loss of visualization in quantum theory, and its partial and temporary recovery at the hands of Schrödinger, see Miller (1984), pp. 125–183. Cultural aspects of the discussion of both the lack of visualization and the indeterminism of quantum theory are emphasized by Forman (1984). Standard histories of quantum physics are Jammer (1966) and Mehra and Rechenberg (1982–1987).

phenomena were visualized as wave effects: diffraction as wave interference, for example. Since the detection of the electron in 1897, subatomic particles were visualized as miniature versions of macroscopic bodies. Particles were attributed the properties of everyday bodies such as billiard balls: they were precisely localized, had a definite mass, velocity, momentum, and kinetic energy, and moved in continuous trajectories. The atomic theory advanced by Rutherford in 1911 visualized the atom as a miniature planetary system in which electrons moved in orbits around the nucleus like planets orbiting the sun.

Around 1900, it became clear that classical physical theory was unable to explain empirical findings about some important submicroscopic phenomena: black-body radiation, the photoelectric effect, and the absorption and emission spectra of atoms. In new theories attempting to explain these phenomena, physicists introduced the notion of a fundamental unit or quantum of energy. The first use of this notion was made by Planck in 1900 in his theory of the spectrum of black-body radiation; Einstein adopted it in his account of 1905 of the rate at which electrons in a metal target are liberated by light in the photoelectric effect; and Bohr in 1913 attempted to explain the lines in the absorption and emission spectra of atoms on the assumption that the energy of electrons in the atom is quantized.

Despite referring to energy quanta, these theories retained many of the visualizations characteristic of previous theories of submicroscopic phenomena. For instance, Bohr's theory of the atom continued to visualize electrons as classical particles, as Rutherford's theory had. The tradition of picturing submicroscopic entities as miniature versions of everyday entities was therefore still unbroken.

These early quantum theories had two shortcomings. First, their empirical success was in some cases modest: for instance, Bohr's theory was found to be incapable of accounting for the behavior of atoms other than the simplest, the hydrogen atom. Second, and more fundamentally, it became increasingly apparent that these theories were insufficiently systematic. Far from providing a unified and coherent account of phenomena, they amounted to a doctoring of classical physical theory: they specified a set of quantum conditions that in certain circumstances had to be imposed upon classical theories for correct results to be obtained.

A much more general and systematic theory of submicroscopic phenomena based on the notion of the quantum was given in 1925 by Heisenberg. This theory, called matrix mechanics, limited itself to relating the magnitudes of observable parameters to one another. Subatomic particles were treated in this theory as abstract entities, whose properties ensured that certain experiments had particular outcomes, but of which no visualization was given. As soon became clear, matrix mechanics gave no visualization of submicroscopic phenomena not merely because Heisenberg had been unwilling to construct one, but because there was none that matrix mechanics could endorse. According to matrix mechanics, quantum particles have no such properties as a precise position, velocity, momentum, or energy. Heisenberg's indeterminacy principle of 1927 stated this clearly. Thus, any attempt to give a pictorial interpretation of submicroscopic phenomena that is consistent with matrix mechanics is thwarted by the lack of macroscopic terms to which the behavior of subatomic particles can be related. Matrix mechanics therefore marked a break in the tradition that theories in submicroscopic physics should offer visualizations of phenomena in macroscopic terms.

Heisenberg asserted that he found the abstractness of matrix mechanics congenial to his style of thinking, which he described as nonvisual. Some other physicists, such as John H. Van Vleck, said that they, too, felt comfortable with the theory's abstractness.[19] But other physicists lamented the loss of visualization. One of these was Schrödinger.

Schrödinger's theorizing was greatly influenced by aesthetic factors. He imposed two aesthetic requirements on physical theories: that the mathematical equations that they contained should have an elegant form and that they should give visualizations of the phenomena that they treated.[20] Guided partly by these principles, and building on de Broglie's idea that waves exhibit particle properties, Schrödinger developed in 1927 a quantum theory of subatomic particles that was named wave mechanics. This theory is (as was soon shown) logically equivalent to Heisenberg's, but seemed to reopen a route to the visualization of submicroscopic phenomena. At the core of the theory was the equation that now bears Schrödinger's name, of which Schrödinger's original (nonrelativistic) formulation is:

$$\nabla^2 \psi + [8\pi^2 m (E - E_{pot})/h^2] \, \psi = 0$$

where ∇^2 is a differential operator, m is the mass of the particle to which the equation applies, E its total energy, and E_{pot} its potential energy. The solutions of this equation are the so-called ψ-functions. Schrödinger interpreted each ψ-function as describing a matter wave with a particular frequency, and on this basis he visualized a subatomic particle as a wave-packet formed by the superposition of a number of such matter waves.

19. Van Vleck (1972), pp. 8–9.
20. Testimony on the importance that Schrödinger attributed to elegance in mathematical equations may be found in Moore (1989), e.g., pp. 196, 384; on his predilection for visualization, see Wessels (1983), pp. 260–273.

Describing the genesis of his theory, Schrödinger made clear that he had striven to find a visualizing alternative to Heisenberg's matrix mechanics: "My theory was inspired by L. de Broglie [. . .] and by short but incomplete remarks by A. Einstein [. . .]. No genetic relation whatever with Heisenberg is known to me. I knew of his theory, of course, but felt discouraged not to say repelled, by the methods of transcendental algebra, which appeared very difficult to me and by the lack of visualizability."[21] The fact that the Schrödinger equation seemed to offer a visualization of submicroscopic phenomena, added to the fact that the mathematics of differential equations was generally more familiar to physicists than that of matrices, inclined most physicists to prefer the theory of wave mechanics to matrix mechanics. Many of the scientists who expressed this preference saw themselves as applying an aesthetic criterion for theory choice.[22] As Jagdish Mehra reports, "The great physicists in Berlin, Planck, Einstein, and M. v. Laue, were very happy with Schrödinger's work because in it one could use continuous functions throughout, and one did not have to rely on the 'nasty and ugly' matrix mechanics."[23] Heisenberg himself acknowledged that wave mechanics possessed some aesthetic appeal, calling it "elegant and simple."[24]

Particularly well documented is Dirac's response to wave mechanics. Dirac saw himself as sharing Schrödinger's aesthetic preferences:

> Of all the physicists that I met, I think Schrödinger was the one that I felt to be most closely similar to myself. I found myself getting into agreement with Schrödinger more readily than with anyone else. I believe the reason for this is that Schrödinger and I both had a very strong appreciation of mathematical beauty, and this appreciation of mathematical beauty dominated all our work. It was a sort of act of faith with us that any equations which describe fundamental laws of Nature must have great mathematical beauty in them. It was like a religion with us. It was a very profitable religion to hold, and can be considered as the basis of much of our success.[25]

Dirac's sympathy for Schrödinger's aesthetic preferences manifested itself in his approbation of wave mechanics. Dirac's well-known account

---

21. Quoted from Miller (1984), p. 143. For fuller details of the rise and reception of matrix and wave mechanics, see Moore (1989), pp. 191–229.
22. Feuer (1957), p. 117, agrees that the choice between wave mechanics and matrix mechanics is made partly on "a criterion of aesthetic elegance."
23. Mehra (1972), p. 35.
24. Heisenberg (1971), pp. 72–73.
25. Dirac (1977), p. 136.

of the development of wave mechanics dwells not only on the role that he thought aesthetic criteria had played in Schrödinger's theorizing but also on the beauty that he himself saw in the theory:

> Heisenberg worked keeping close to the experimental evidence about spectra [. . .]. Schrödinger worked from a more mathematical point of view, trying to find a beautiful theory for describing atomic events [. . .]. He was able to extend De Broglie's ideas and to get a very beautiful equation, known as Schrödinger's wave equation, for describing atomic processes. Schrödinger got this equation by pure thought, looking for some beautiful generalization of De Broglie's ideas, and not by keeping close to the experimental development of the subject in the way Heisenberg did.[26]

Dirac approved not only of the elegance of Schrödinger's equation but also of the theory's power of visualization: he said that he liked it that wave mechanics visualized particles as variations in the density of some medium spread out in space.[27]

In the longer term, however, Schrödinger's wave mechanics did not succeed in providing a consistent visualization of submicroscopic phenomena in classical terms. The attempt foundered above all on the empirical evidence that, while some properties of subatomic particles resemble those of classical waves, other properties of theirs resemble those of classical particles. An example of the difficulty of visualizing the behavior of quantum particles is given by Bohr:

> The extent to which renunciation of the visualization of atomic phenomena is imposed upon us [. . .] is strikingly illustrated by the following example to which Einstein very early called attention and often has reverted. If a semi-reflecting mirror is placed in the way of a photon, leaving two possibilities for its direction of propagation, the photon may either be recorded on one, and only one, of two photographic plates situated at great distances in the two directions in question, or else we may, by replacing the plates by mirrors, observe effects exhibiting an interference between the two reflected wave-trains. In any attempt of a pictorial representation of the behaviour of the photon we would, thus, meet with the difficulty: to be obliged to say, on the one hand, that the

26. Dirac (1963), pp. 46–47.
27. On Dirac's approval of wave mechanics, see Mehra (1972), pp. 50–51, and Kragh (1990), pp. 30–37.

photon always chooses *one* of the two ways and, on the other hand, that it behaves as if it had passed *both* ways.[28]

To escape from this conundrum one must cease trying to determine to what extent subatomic particles and electromagnetic radiation are waves or particles that are visualizable in classical terms, and conceive of them instead as wave-particles or "wavicles," entities unknown to classical physical theory and to everyday experience. This view goes under the name of wave–particle dualism.[29] On the strength of these arguments, Schrödinger's visualization of ψ-functions in terms of waves was rejected by most physicists, in favor of a statistical interpretation that did not lend itself to any visualization. The so-called Copenhagen interpretation of quantum theory, which became the majority view of the physics community in the 1930s, maintained this statistical reading.[30]

The property of quantum theory that it offers no convincing visualization of subatomic particles evoked two responses among physicists after 1928. One group never reconciled itself to the fact that physical theory had become unvisualizing. Schrödinger's own later career exemplifies the attitude of these physicists. He continued to maintain that visualization was an essential property of a good theory. Rather than repudiate this requirement, he sought areas of physics to which his style of theorizing was better suited: for the following twenty-five years he largely abandoned quantum mechanics for such areas as relativity physics. In the 1950s, toward the end of his life, he wrote a series of papers calling for visualization in quantum mechanics and proposing a new version of his wave interpretation of the ψ-functions; but these suggestions were not consonant with the state of the discipline and were not taken up.[31] Many physicists of this group were able to distinguish the abstractness of quantum theory, which they deplored, from its empirical success, of which they had a high opinion; but they were not persuaded that the latter property compensated for the former.

A second group of physicists, led by Bohr and Heisenberg, judged the loss of visualization a price worth paying for an empirically successful theory of submicroscopic phenomena. They continued to produce extensions and refinements of quantum theory in the abstract style. Heisenberg's stance was, as he later explained, the following:

28. Bohr (1949), p. 222.
29. The rise of wave–particle dualism is retraced by Wheaton (1983).
30. On physicists' successive interpretations of quantum theory, see Jammer (1974).
31. On Schrödinger's failure to re-establish visualization in quantum theory, see Wessels (1983), pp. 265–269; for information on his later career, see Moore (1989).

Continuity and Revolution in Twentieth-Century Physics

Classical physics has taught us to talk about particles and waves, but since classical physics is not true there, why should we stick so much to these concepts? Why should we not simply say that we cannot use these concepts with a very high precision, therefore the uncertainty relations, and therefore we have to abandon these concepts to a certain extent. When we get beyond this range of the classical theory, we must realize that our words don't fit. They don't really get a hold in the physical reality and therefore a new mathematical scheme is just as good as anything because the new mathematical scheme then tells what may be there and what may not be there.[32]

Even Dirac, who had appreciated Schrödinger's attempt to save visualization, continued to contribute to quantum theory in the abstract style inaugurated by Heisenberg, not letting the loss of visualization deprive him of an empirically successful theory.

Physicists of this second group were not entirely unperturbed at quantum theory's lack of visualization: many of them, like members of the first group, found it unsettling not to be able to trust their visual intuitions in developing and applying physical theory. The disquiet at the loss of visualization found expression in utterances about quantum theory of an incredulous, sarcastic, or paradoxical tone. One example is the quip that if you understand quantum theory, then you do not understand it; another is the joke, which seems to have first been made in 1928 by William H. Bragg, that electrons and photons behave as waves on Mondays, Wednesdays, and Fridays and as particles on Tuesdays, Thursdays, and Saturdays.[33]

These two groups of physicists exemplify the conservative and progressive factions that form in a community undergoing a scientific revolution. The commitment to visualization felt by physicists of the first group was induced by the long experience of seeing empirically successful theories offer visualizations of their subject matter: it was a product of the aesthetic induction operating over the record of classical physics. Physicists of the second group were able to embrace quantum theory by relaxing their commitment to the requirements of classical physics. For instance, Bohr wrote that quantum theory was made possible "only by a conscious resignation of our usual demands for visualization and causality."[34]

As a result of quantum theory's continuing empirical success, the

32. Quoted from Pais (1991), p. 310.
33. Wheaton (1983), p. 306.
34. Bohr (1934), p. 108.

physics community increasingly dissociated itself from its earlier requirement that theories should offer visualizations of submicroscopic phenomena. In the 1930s, physicists loosened the requirement of visualization reluctantly, in response to the seeming impossibility of satisfying it. With the passage of time, however, many physicists began to express regret that this requirement ever carried the weight that it did, and some even portrayed it as having hindered the development of physics. The following passage by John Gribbin, written in 1984, displays this new attitude.

> Without doubt, the appealing picture of physically real waves circling around atomic nuclei that had led Schrödinger to discover the wave equation [. . .] is wrong. Wave mechanics is no more a guide to the reality of the atomic world than matrix mechanics, but unlike matrix mechanics wave mechanics gives an *illusion* of something familiar and comfortable. It is that cozy illusion that has persisted to the present day and that has disguised the fact that the atomic world is totally different from the everyday world. Several generations of students, who have now grown up to be professors themselves, might have achieved a much deeper understanding of quantum theory if they had been forced to come to grips with the abstract nature of Dirac's approach, rather than being able to imagine that what they knew about the behavior of waves in the everyday world gave a picture of the way atoms behave.[35]

The change in attitude illustrates how, in the wake of a revolution, scientists reassess the merits of their aesthetic commitments. As the empirical track record of quantum theory improved, its aesthetic properties reshaped the aesthetic canon of physicists. Since quantum theory is an abstract theory, the canon has come increasingly to value abstractness and penalize visualization. Gribbin's denunciation of the requirement of visualization, which would have been unthinkable in the 1920s, shows how far this amendment of aesthetic canons has proceeded.

4. THE RENUNCIATION OF DETERMINISM

Quantum theory's metaphysical implications aggravated the misgivings that its abstractness evoked. Some of the implications of quantum theory conflicted sharply with long-held metaphysical presuppositions. For example, the claim that the energy and other properties of physical systems

35. Gribbin (1984), p. 117.

come in discrete units rather than continuously varying quantities conflicts with the principle of the continuity of nature. This principle, often expressed in the dictum *Natura non facit saltus* ("Nature makes no leaps"), can be retraced to Aristotle and was regarded by Leibniz as a fundamental tenet of natural philosophy.[36] Physical theories from the time of Galileo to the late nineteenth century had invariably been consistent with this principle, so in 1900 physicists associated all their empirical successes with theories that portrayed nature as continuous. When quantum theory arose, some observers argued that the fact that it portrayed nature as discontinuous showed it to be wrong, on the grounds that the success of classical physical theories had established that nature abhorred discontinuities. The preference for continuity can be described as the result of the aesthetic induction operating over the record of classical physics; and the distaste for theories that portrayed nature as discontinuous can easily be seen to have been partly aesthetic. Leibniz, for example, expressed aesthetic appreciation for physical theories that portrayed nature as continuous.[37]

In this section, however, we focus on another of the distinctive metaphysical properties of quantum theory, its indeterminism. A deterministic theory is one that portrays a physical system as being such that it is possible, from knowledge of an initial state of it, to predict a future state of it to similar degree of detail. The theories of classical physics are deterministic. By contrast, an indeterministic theory is one that portrays a physical system as not allowing, even in principle, such predictions.

The fact that quantum theory is indeterministic became apparent only gradually. It was not yet obvious when Planck formulated the quantum theory of black-body radiation in 1900, when Einstein explained the photoelectric effect on quantum principles in 1905, or even when Bohr formulated his quantum theory of the atom in 1913. It was recognized fully only in the so-called new quantum theory—developed from about 1925 onwards by Bohr, Heisenberg, Schrödinger, and others—of which both matrix and wave mechanics are components.

Many physicists who had contributed to the early development of quantum ideas found the indeterminism of the new quantum theory unpalatable. Representative of such scientists are Planck and Einstein. Neither denied that the new quantum theory had considerable empirical worth: indeed, both repeatedly paid tribute to its predictive success. Their dissatisfaction was motivated by properties of the theory other

36. The history of the principle of continuity is retraced in Lovejoy (1936).
37. Leibniz's attribution of aesthetic value to the principle of continuity is documented in Breger (1994), pp. 133–135.

than its empirical capabilities, and chiefly by its indeterminism. In their overall assessment of the theory, this property easily outweighed its empirical success and ensured their rejection of it.[38]

In his Nobel Prize address, in which he retraces the development of quantum theory, Planck explicitly and at length commends the theory for its empirical success in many areas of physics.[39] Nonetheless, he shows his displeasure with it:

> The difficulties which the introduction of the quantum of action into the well-established classical theory has encountered from the outset [. . .] have gradually increased rather than diminished; and although research in its forward march has in the meantime passed over some of them, the remaining gaps in the theory are the more distressing to the conscientious theoretical physicist. [. . .]
>
> But numbers decide, and in consequence the tables have been turned.[40]

Planck's expression of dissatisfaction with quantum theory indicates that he harbored extraempirical reservations about it, irrespective of its empirical success, which, he admitted, had won over the scientific community.

Einstein took a similar attitude. He felt that if physics since Newton had achieved so many triumphs, it was because the discipline had recognized the value of formulating deterministic theories. He believed that physicists ought to persist in formulating such theories, even in the domain of submicroscopic phenomena:

> What has happened since Newton in theoretical physics is the organic development of his ideas. Force became independent reality to Faraday, Maxwell, and Lorentz, and then went over into the conception of the field. The partial differential equation has taken the place of the ordinary differential equation used by Newton to express causality. Newton's absolute and fixed space has been converted by the theory of relativity into a physically vital frame. It is only in the quantum theory that Newton's differential method becomes inadequate, and indeed strict causality fails

38. For further information on Planck's resistance to quantum theory, see Heilbron (1986), pp. 122–140; on Einstein's, see Stachel (1986) and Ben-Menahem (1993). In contrast, Schrödinger was not greatly troubled by the theory's indeterminism, as Ben-Menahem (1989) documents.

39. Planck (1922), pp. 13–17.

40. Ibid., p. 18. Planck recalls his opposition to quantum theory in his (1948), pp. 43–45.

us. But the last word has not yet been said. May the spirit of Newton's method give us the power to restore unison between physical reality and the profoundest characteristic of Newton's teaching—strict causality.[41]

To help ensure this outcome, in the 1920s Einstein began a long campaign to cast doubt on quantum theory and especially on Heisenberg's indeterminacy relations, which describe limits to the precision with which certain physical quantities are determined. At first, he argued that quantum theory is internally inconsistent; about 1935 he switched to arguing that the theory is an incomplete representation of physical reality. Much of this campaign was conducted in debates with Bohr.[42]

To try to show inconsistencies in quantum theory, Einstein would propose thought experiments in which, he claimed, physical quantities could be measured more precisely than the indeterminacy relations allow. His best-known thought experiment envisaged a box containing a source of radiation and suspended from a spring balance. A shutter in the wall of the box is operated by a clock within the box. At some instant the clock opens the shutter briefly, allowing one photon of radiation to escape from the box. The decrease in the energy of the box can be measured on the balance as a decrease in mass, and the time of the photon's escape can be measured by the clock. Einstein claimed that these two quantities can be measured to arbitrary accuracy, in violation of the indeterminacy relations. In reply, Bohr showed that, on the general theory of relativity, the change in position of the box in the gravitational field introduces uncertainties in the measurements of energy and time, which accord with Heisenberg's relations.

By reconciling Einstein's thought experiments with quantum theory, Bohr showed that no inconsistencies had been found in the theory. That Einstein's opposition was not allayed by these exchanges suggests that it was based on other concerns, especially his commitment to determinism. Einstein's choice to pursue his campaign against quantum theory by means of consistency arguments was motivated probably by the feeling that they would carry more weight in the scientific community than openly metaphysical reasoning. After all, everyone recognized the need to avoid logical inconsistencies, but few of the scientists involved in developing quantum theory shared Einstein's commitment to determinism.

Einstein's arguments in the second phase of his campaign, directed at establishing that quantum theory is an incomplete representation of reality, reflect the distinction between the empirical worth of the theory and its acceptability on other grounds:

41. Quoted from *Nature* (1927), p. 467.
42. Some references to studies of the Bohr–Einstein debate were given in Chapter 8.

Experiments on interference made with particle rays have given a brilliant proof that the wave character of phenomena of motion as assumed by the theory does, really, correspond to the facts. In addition to this, the theory succeeded, easily, in demonstrating the statistical laws of the transition of a system from one quantum condition to another under the action of external forces, which, from the standpoint of classical mechanics, appears as a miracle. [. . .] Even an understanding of the laws of radioactive decomposition, at least in their broad lines, was provided by the theory.

Probably never before has a theory been evolved which has given a key to the interpretation and calculation of such a heterogeneous group of phenomena of experience as has the quantum theory. In spite of this, however, I believe that the theory is apt to beguile us into error on our search for a uniform basis for physics, because, in my belief, it is an *incomplete* representation of real things [. . .]. The incompleteness of the representation is the outcome of the statistical nature [. . .] of the laws.[43]

Like Planck's, Einstein's stance toward quantum theory—recognition of its empirical worth but overall rejection of it on the grounds that it fails to meet certain requirements—is reminiscent of the behavior of conservative scientists in a revolution (Chapter 8). Indeed, I suggest that Planck and Einstein are best seen as conservatives confronting a revolutionary innovation.

If I am to attribute such a role to Planck and Einstein, I must interpret their rejection of quantum theory as grounded on aesthetic criteria. It is not difficult to find support for this suggestion. First, determinism and indeterminism are metaphysical doctrines, and I have argued that the metaphysical allegiances of theories should be counted among their aesthetic properties. Second, Einstein's biographers concur that his misgivings about indeterminism arose from an aesthetic feeling: for him the harmony of the theory would be marred if, to use his own metaphor, it depicted God as deciding occurrences on the cast of a die.[44]

For my model of scientific practice to apply fully to Planck's and Einstein's commitment to determinism, I must furthermore interpret this commitment as formed by the aesthetic induction. I think this suggestion too is supported by the evidence. In urging their colleagues to adhere to the style of theorizing developed by classical physics, Planck and Einstein were expressing a conviction that the magnificent empirical track

43. Einstein (1936), p. 374.
44. Einstein's rejection of indeterminism is portrayed as an aesthetic response by Hoffmann and Dukas (1972), pp. 193–195.

record of classical physical theories had to be related to their determinism. They were thereby performing the aesthetic induction over the empirical performance of theories showing the property of determinism.

Bohr, in contrast, belonged to the progressive faction in the revolution. As we have seen, members of such a faction tend to suspend allegiance to any aesthetic preferences and conduct theory choice exclusively on empirical criteria. There is good evidence that Bohr had little commitment to an aesthetic canon. As Léon Rosenfeld writes of him, "In speculating about the prospects of some line of investigation, he would dismiss the usual considerations of simplicity, elegance or even consistency with the remark that such qualities can only be properly judged after the event: 'I cannot understand,' he used to say, 'what it means to call a theory beautiful if it is not true.' "[45] This phrase implies an unwillingness to talk about the beauty of theories until after their empirical worth has become clear. Bohr accepts only logical and empirical criteria as the grounds for theory evaluation: "In my opinion, there could be no other way to deem a logically consistent mathematical formalism as inadequate than by demonstrating the departure of its consequences from experience or by proving that its predictions did not exhaust the possibilities of observation."[46] This emphasis on the empirical acceptability of theories amounts to a form of positivism, and Bohr was indeed generally taken by his contemporaries as a positivist in these matters.[47]

The debates between Einstein and Bohr thus illustrate the disagreement between the conservative and progressive factions in a scientific revolution. While the conservative faction wishes to hold to established aesthetic criteria in theory choice, the progressive faction relaxes all extraempirical constraints on theorizing. These attitudes are captured precisely in the way Bohr characterizes Einstein and himself:

> Notwithstanding the most suggestive confirmation of the soundness and wide scope of the quantum-mechanical way of description, Einstein [. . .] expressed a feeling of disquietude as regards the apparent lack of firmly laid down principles for the explanation of nature, in which all could agree. From my viewpoint, however, I could only answer that, in dealing with the task of bringing order into an entirely new field of experience, we could hardly trust in any accustomed principles, however broad, apart from the demand of avoiding logical inconsistencies and, in

45. Rosenfeld (1967), p. 117. Mott (1986), p. 25, writes nonetheless that he learned from Bohr "how beautiful physics could be."
46. Bohr (1949), p. 229.
47. On the perception of Bohr as a positivist, see Murdoch (1987), pp. 139–140.

this respect, the mathematical formalism of quantum mechanics should surely meet all requirements.[48]

Under the operation of the aesthetic induction, and on the strength of the empirical success of quantum theory, we might expect physicists since the 1920s to have developed an increasing attraction for indeterminism, just as they have come to accept abstractness. This has indeed occurred. Few physicists today remain opposed to quantum theory's indeterminism, and a new theory's indeterminism no longer counts against its adoption. Moreover, quantum theory has gradually come to be seen as aesthetically pleasing. Looking back in 1970, Heisenberg admits that at first there was a sense of aesthetic loss. "Since Planck's discovery of the quantum of action, in 1900, a state of confusion had arisen in physics. The old rules, whereby nature had been successfully described for more than two centuries, would no longer fit the new findings. [. . .] The beauty and completeness of the old physics seemed destroyed, without anyone having been able, from the often disparate experiments, to gain a real insight into new and different sorts of connection." But when the internal harmony of a science is lost, it is generally recovered once the foundations of the subject have been reconstituted. "In atomic physics this process took place not quite fifty years ago, and has again restored exact science, under entirely new presuppositions, to that state of harmonious completeness which for a quarter of a century it had lost."[49] At around the time that Heisenberg wrote this passage, Max Jammer felt able to appraise quantum theory as "an imposing intellectual structure of great beauty."[50] The discrepancy between these statements and Planck's and Einstein's declarations of revulsion at quantum theory is a demonstration of the power of the aesthetic induction to foster aesthetic appreciation for empirically successful theories.

48. Bohr (1949), p. 228.
49. Heisenberg (1970), pp. 181, 182.
50. Jammer (1966), p. v.

· C H A P T E R   T W E L V E ·

Rational Reasons
for Aesthetic Choices

1. REVIEW OF RESULTS

The defense of the rationalist image of science presented in this book is now complete. Before drawing some wider conclusions, let us review its principal steps.

The chief claim of the rationalist image of science is that there exists a set of precepts for conducting science—the norms of rationality—for which a principled and extrahistorical justification can be given. The rationalist image is also committed to portraying actual scientific practice as complying to a large extent with these precepts. The rationalist image would therefore come under question if it were established that scientists' reasoning and decisions departed substantially from the norms of rationality. Many philosophers and historians of science think that two bodies of historical evidence show that scientists' behavior does indeed depart substantially from rationality. The first is the evidence that scientists choose among alternative theories partly on aesthetic criteria; the second is the evidence that scientific practice undergoes revolutions, in which the criteria that scientists use to evaluate theories change radically. The aim of this book has been to construct accounts of these phenomena that are consistent with the rationalist image.

Two suggestions that cannot be maintained are that scientists pass aesthetic judgments on their theories in an attitude of disinterestedness toward empirical success and that the aesthetic judgment that a scientist passes on a theory is an aspect of his or her empirical evaluation of it. Evidence shows that although scientific communities tend to attach aesthetic value to properties of theories that have demonstrated empirical success, aesthetic appraisals do not invariably concur with appraisals on empirical criteria. According to my model of scientists' aesthetic evaluations of theories, aesthetic canons are formulated and updated by scientific communities by means of a mechanism that I have called the aesthetic induction. When examining the empirical record of theories in their discipline, scientists attach to each aesthetic property a weighting roughly proportional to the degree of empirical adequacy that they attribute to theories exhibiting that property. The table of weightings constructed in this way constitutes the scientists' aesthetic canon, used thereafter to evaluate theories in their discipline.

It may be that some aesthetic properties exist that are correlated with high degrees of empirical adequacy. A scientific realist would describe this eventuality as one in which all theories that are within a certain distance from the true theory of the universe have particular aesthetic properties. If such aesthetic properties existed, it would be possible to formulate criteria for theory choice that enjoined scientists to prefer theories that possess them. Though these criteria would refer to aesthetic properties of theories, they would be suited to furthering the same aims of theory choice to which empirical criteria are directed: they would enable scientists presented with a choice between two theories to diagnose which had the greater degree of empirical adequacy.

If such aesthetic properties existed, persistent use of the aesthetic induction would reveal which they were: they would receive an ever-increasing weighting in scientists' aesthetic canons. However, we have as yet no convincing evidence that such aesthetic properties exist. On the contrary, most aesthetic preferences that scientists have hitherto conceived have eventually proved a hindrance to the pursuit of empirical success. Indeed, scientists who have wished to adopt the empirically most successful theories available at each time have found it necessary periodically to repudiate established aesthetic preferences. In the light of this fact, the claim endorsed by Einstein, Dirac, and others—that there are aesthetic criteria already in use that are reliable indicators of theories' degrees of empirical adequacy—cannot be sustained.

The events in which scientists relax their commitment to established aesthetic canons are scientific revolutions. One consequence of a revolution is that the set of aesthetic criteria governing theory choice alters. The theories that are formulated and adopted after a revolution differ in aesthetic properties from those before. Nonetheless, there is a partial continuity of scientific practice before and after, assured by the continuity of the community's empirical criteria. Since a scientific revolution occasions only a partial change of criteria for the evaluation of theories, the reasoning and deliberations of scientists before a revolution remain partially comprehensible to scientists and historians afterward.

2. A Rational Warrant for Aesthetic Commitments

Now that we are equipped with a model of the origin and evolution of aesthetic canons, we are able to reassess the problem posed to the rationalist image of science by the fact that scientists rely on aesthetic criteria in choosing among theories. Should we consider it a serious departure from rationality for scientists to allow their judgments of theories to be determined partly by aesthetic criteria?

A simple-minded analysis suggests that it is indeed contrary to rationality to use the aesthetic induction as a source of criteria for theory choice. The goal of science is the production of the most complete and accurate account possible of the universe. Our understanding of which properties are conducive to theories' having high degrees of empirical adequacy is provided by goal analysis, which yields our empirical criteria for theory assessment. To allow our judgments of theories to depart from the verdicts delivered by these empirical criteria is thus to deviate from what it is rational to do. Although the verdicts delivered by aesthetic criteria do reflect the past empirical performance of theories, we cannot be assured that they will agree with the verdicts of empirical criteria. In cases where the verdicts of aesthetic criteria concur with those of empirical criteria, their use brings no advantage; in other cases, aesthetic criteria may lead the community to choose theories that are empirically less successful. Thus, to allow our appraisals of theories to be affected by aesthetic criteria is to depart from rationality. On this view, actual scientific practice indeed contradicts the rationalist image of science, since scientists' appraisals of theories are often determined in part by aesthetic criteria.

A more sophisticated view of the aesthetic induction, however, suggests that it is rationally justifiable to allow our appraisals of theories to be shaped in part by aesthetic criteria. There may exist certain aesthetic properties—certain simplicity properties, certain symmetry properties, and so on—that are conducive to theories' having high degrees of empirical adequacy. This is the possibility that we uncovered in Chapter 6. The pragmatic justification of inductive policies assures us that, if such aesthetic properties of theories exist, the inductive projection will be at least as likely to discover them as any alternative procedure for formulating criteria. Whether the strategy consisting of the aesthetic induction is rationally justified depends on the probabilities and the payoffs of its possible outcomes. The possible outcomes are two: the aesthetic induction may identify aesthetic properties that are conducive to theories' having high degrees of empirical adequacy, and it may fail to identify any such properties.

The payoffs of these outcomes are not difficult to evaluate. As long as it identifies no aesthetic properties of theories that are correlated with high degrees of empirical adequacy, the aesthetic induction is somewhat detrimental to empirical performance, since it leads us on some occasions to make empirically suboptimal choices among theories. These disadvantageous choices are redressed in scientific revolutions, when aesthetic preferences that have been demonstrated to hamper the pursuit of empirical success are abandoned. By contrast, the discovery of aesthetic properties of theories that are correlated with high degrees of empirical adequacy would bring considerable benefits. Such a discovery would enlarge our set of criteria for recognizing scientific theories that are likely to be empirically successful. More importantly, it would reveal new facets of the concepts of truth and beauty, transforming epistemology and aesthetics.

Now for the probability of these outcomes. Estimates of the likelihood that the aesthetic induction will identify aesthetic properties that are conducive to theories' having high degrees of empirical adequacy are probably no more trustworthy than guesses. Our estimates will be decided partly by our attitude toward the claim that the perceptual features of entities accord with their practical qualities. Those who hold to versions of Platonism or Pythagoreanism generally assert the existence of such an accord; their ranks have included Einstein and Dirac. Others reject this claim vehemently.

In view of the difficulty of ascertaining the likelihood of the aesthetic induction's success, the most reasonable conclusion may be that, as long as we cannot rule out that aesthetic properties exist which are conducive to theories' having high degrees of empirical adequacy, it would be foolish to deny ourselves the chance of discovering them. We should therefore continue to perform the aesthetic induction. According to this conclusion, appeal to aesthetic criteria in theory choice is rationally justifiable, and it is therefore compatible with the rationalist image of science.

3. THE RATIONALITY OF REVOLUTIONS

One of the central questions in discussions of scientific revolutions has been whether undertaking a revolution can be a rational act or is invariably nonrational. A scientific revolution involves two changes: the transfer of a community's allegiance from one theory to another with radically different properties, and a change in the community's set of criteria for theory choice. In order for it to be rational to undertake a revolution, there must exist some criteria for comparing the theory that is relinquished with the one that replaces it and for judging the latter to be superior. If there are no such criteria, the act of undertaking a revolution must be regarded as nonrational: on this view, while there are causes of revolutions, there are no reasons for them.

Some models of scientific practice portray revolutions as consisting of a replacement of a community's set of criteria for theory choice in its entirety by a new set. Such a model is that of Kuhn, if the radical reading of him is correct. All such models entail that undertaking a revolution is nonrational, as the following argument shows. To compare the worth of the theory that was relinquished during a revolution, T_1, and the theory that was accepted in its place, T_2, a set of criteria is needed. Only two sets of criteria could legitimately be used for this task: the set that was relinquished during the revolution, C_1, and the one that was accepted in its place, C_2. Any other set of criteria that might be envisaged would not have been recognized by the participants in the revolution and would thus not shed light on whether the revolution was justified. Presumably, C_2 recommends replacing T_1 by T_2, while C_1 recommends not doing so: otherwise, no revolution would have occurred. So if we are to deem the replacement of T_1 by T_2 to have been justified, we must show the replacement of C_1 by C_2 to have been justified. In order to show this, we need a set of criteria. But C_1 and C_2 exhaust the criteria that would have been recognized by the participants in the revolution. This means that we cannot show either the replacement of C_1 by C_2 or, consequently, that of T_1 by T_2 to have been justified. Thus, undertaking the revolution was a nonrational act. As Kuhn would put it, if each criterion for theory choice holds only within a particular paradigm, there can be no rational grounds on which to choose between alternative paradigms.

In contrast, models that portray revolutions as a change in less than the community's entire set of criteria for theory choice are able to deem some revolutions to be rational. This is because the criteria for theory choice that endure unchanged through the revolution offer a basis recognized by all participants for adjudicating between theories. If this set of criteria rates T_2 above T_1, the replacement of T_1 by T_2 was justified, and there was a rational justification for undertaking this revolution; if not, then both the replacement and the revolution were unjustified. The model of scientific practice that I have presented takes this form. While each revolution involves a change in a community's set of aesthetic criteria, its set of empirical criteria survives successive revolutions unchanged. On this model, therefore, there can be rational grounds for undertaking revolutions. This is what the progressive faction in a revolution typically argues in advocating that an established aesthetic canon should be abandoned: its members believe that the progress of science will be better promoted if the constraints on theory choice imposed by the aesthetic canon are relaxed.

In summary, there may be a rational justification both for choosing theories on established aesthetic criteria and for abandoning those criteria in revolutions. As long as there endures a correlation between a theory's showing particular aesthetic properties and its demonstrating empirical success, it may be rational to continue choosing on those aesthetic criteria. After all, this policy will reveal any genuine link that may exist between the empirical success of theories and their aesthetic properties. On the other hand, if a correlation between a theory's showing particular aesthetic properties and its demonstrating empirical success breaks down, such that continuing to favor the theory requires a sacrifice of empirical performance, it is rational to abandon those aesthetic preferences in a revolution.

4. A NATURAL INDUCTIVE DISPOSITION

I have been speaking as though it were open to scientific communities and individual scientists to decide whether to perform the aesthetic induction. If this were so, then if scientists judged that it is irrational to allow their theory choices to be affected by aesthetic criteria, they would be able to refrain from doing so, and choose among theories on nothing other than empirical criteria.

In fact, I believe that scientists and scientific communities do not generally have the capability to carry out such a decision. I believe that I have identified a natural phenomenon of scientific communities: a disposition to associate aesthetic properties of theories with expectations of empirical success and to conduct choices among theories in the light of these expectations. This tendency is largely involuntary: scientists are mostly unable to prevent themselves from either forming these expectations or conducting theory choice in consequence of them. While it may be possible to bring scientists to recognize in general terms that a strong commitment to particular aesthetic properties of theories may hamper the pursuit of empirical success, I would not expect this recognition greatly to affect their behavior in theory choice. In any specific instance of theory choice, scientists will regard their current aesthetic predilections—whichever they are—as natural and proper. Moreover, since scientists' aesthetic canons are formed by the aesthetic induction, any scientist will be able at any time to defend his or her aesthetic predilections by pointing to a past correlation between empirical success and whichever aesthetic property of theories he or she values. Individual scientists are therefore always able to portray themselves as having good reasons for holding to their particular aesthetic predilections.

Only a few scientists at certain junctures succeed in suspending aesthetic commitments and escaping from the aesthetic induction. For scientists to be brought to this step, they must believe both that the established aesthetic preferences are impeding the adoption of theories that have greater empirical worth and that these preferences have no separate justification. The outcome, of course, is a scientific revolution.

As an illustration of the involuntary nature of the aesthetic induction, consider again the responses of Planck, Einstein, and Schrödinger toward quantum theory. Manifestly, Planck's and Einstein's predilection for determinism and Schrödinger's for visualization prevented them from embracing the empirically best-performing theories of submicroscopic phenomena that were available to them. I conjecture that if their aesthetic preferences had been explicitly challenged, however, they would have portrayed them as natural and proper in physics and defended them by pointing to the long sequence of empirically successful theories that had been deterministic and visualizing. I am not confident that an understanding of the way in which aesthetic preferences become entrenched in scientific communities would have made them examine their aesthetic predilections more critically.

In short, I take a Humean view of the disposition of scientists and scientific communities to link theories' aesthetic properties with empirical performance. Hume regarded the formation of beliefs by induction as the manifestation of a tendency of persons to expect any observed associations between occurrences to endure. According to Hume, for example, someone for whom the sight of fire is associated with pain has a tendency to recoil from any subsequent instance of fire.[1] In the aesthetic induction, analogously, scientists for whom certain aesthetic properties are associated with empirical success will value and pursue other theories exhibiting the same properties. Hume did not believe that those who read his treatise would cease making inductive generalizations; similarly, I do not expect my readers to escape the influence of the aesthetic induction.

1. Hume (1739), pp. 98–106.

References

Ackerman, James. 1985. "The Involvement of Artists in Renaissance Science." In John W. Shirley and F. David Hoeniger, eds., *Science and the Arts in the Renaissance*. Washington, D.C.: Folger Books; pp. 94–129.

Agassi, Joseph. 1964. "The Nature of Scientific Problems and Their Roots in Metaphysics." In Mario Bunge, ed., *The Critical Approach to Science and Philosophy*. New York: Free Press; pp. 189–211.

Alexenberg, M. 1981. *Aesthetic Experience in Creative Process*. Ramat Gan: Bar-Ilan University Press.

Aramaki, Seiya. 1987. "Formation of the Renormalization Theory in Quantum Electrodynamics." *Historia Scientiarum* 32:1–42.

Ariew, Roger. 1987. "The Phases of Venus before 1610." *Studies in History and Philosophy of Science* 18:81–92.

Arnheim, Rudolph. 1969. *Visual Thinking*. Berkeley: University of California Press.

Arregui, Jorge V., and Pablo Arnau. 1994. "Shaftesbury: Father or Critic of Modern Aesthetics?" *British Journal of Aesthetics* 34:350–362.

Aune, Bruce. 1977. *Reason and Action*. Dordrecht: Reidel.

Bachelard, Gaston. 1934. *The New Scientific Spirit*. Translated by Arthur Goldhammer. Boston: Beacon Press, 1984.

Badash, Lawrence. 1987. "Ernest Rutherford and Theoretical Physics." In Kargon and Achinstein (1987), pp. 349–373.

Banham, Reyner. 1960. *Theory and Design in the First Machine Age*. 2d ed., 1962. London: Butterworth Architecture.

Barker, Peter. 1981. "Einstein's Later Philosophy of Science." In Peter Barker and Cecil G. Shugart, eds., *After Einstein: Proceedings of the Einstein Centennial Celebration at Memphis State University*. Memphis, Tenn.: Memphis State University Press; pp. 133–145.

Barker, Stephen F. 1957. *Induction and Hypothesis: A Study of the Logic of Confirmation*. Ithaca: Cornell University Press.

Barrow, John D. 1988. *The World within the World*. Oxford: Clarendon Press.

Barthes, Roland, and André Martin. 1964. *La Tour Eiffel*. Paris: Delpire.

Beardsley, Monroe C. 1973. "What Is an Aesthetic Quality?" *Theoria* 39:50–70.

Beer, Gillian. 1983. *Darwin's Plots: Evolutionary Narrative in Darwin, George Eliot and Nineteenth-Century Fiction*. London: Routledge and Kegan Paul.

Bellone, Enrico. 1973. *I modelli e la concezione del mondo nella fisica moderna da Laplace a Bohr*. Milan: Feltrinelli.

Ben-Menahem, Yemima. 1989. "Struggling with Causality: Schrödinger's Case." *Studies in History and Philosophy of Science* 20:307–334.

——. 1993. "Struggling with Causality: Einstein's Case." *Science in Context* 6:291–310.

Bergmann, Peter G. 1982. "The Quest for Unity: General Relativity and Unitary Field Theories." In Holton and Elkana (1982), pp. 27–38.

Bernstein, Jeremy. 1979. *Experiencing Science*. London: Burnett.

Billington, David P. 1983. *The Tower and the Bridge: The New Art of Structural Engineering*. New York: Basic Books.

Blackburn, Simon. 1984. *Spreading the Word: Groundings in the Philosophy of Language*. Oxford: Clarendon Press.

——. 1985. "Errors and the Phenomenology of Value." In Ted Honderich, ed., *Morality and Objectivity*. London: Routledge and Kegan Paul; pp. 1–22.

Bohr, Niels. 1934. *Atomic Theory and the Description of Nature*. Cambridge: Cambridge University Press.

——. 1949. "Discussion with Einstein on Epistemological Problems in Atomic Physics." In Schilpp (1949), pp. 199–241.

Boltzmann, Ludwig. 1901. "The Recent Development of Method in Theoretical Physics." Translated by Thomas J. McCormack. *The Monist* 11:226–257.

Boullart, Karel. 1983. "Mathematical Beauty as a Metaphysical Concept: The Aesthetics of Rationalism." In G. W. Leibniz-Gesellschaft, ed., *Leibniz: Werk und Wirkung. IV. Internationaler Leibniz-Kongreß: Vorträge*. Hannover: G. W. Leibniz-Gesellschaft; pp. 69–76.

Brackenridge, J. Bruce. 1982. "Kepler, Elliptical Orbits, and Celestial Circularity: A Study in the Persistence of Metaphysical Commitment," parts 1 and 2. *Annals of Science* 39:117–143, 265–295.

Braithwaite, Richard B. 1953. *Scientific Explanation: A Study of the Function of Theory, Probability and Law in Science*. Cambridge: Cambridge University Press.

Breger, Herbert. 1989. "Symmetry in Leibnizean Physics." In *The Leibniz Renaissance*. Florence: Leo S. Olschki; pp. 23–42.

——. 1994. "Die mathematisch-physikalische Schönheit bei Leibniz." *Revue internationale de philosophie* 48:127–140.

Brown, Laurie M., and Helmut Rechenberg. 1987. "Paul Dirac and Werner Heisenberg—A Partnership in Science." In Kursunoglu and Wigner (1987), pp. 117–162.

Buchdahl, Gerd. 1970. "History of Science and Criteria of Choice." In Stuewer (1970), pp. 204–230.

——. 1973. "Explanation and Gravity." In Mikuláš Teich and Robert Young, eds., *Changing Perspectives in the History of Science: Essays in Honour of Joseph Needham*. London: Heinemann; pp. 167–203.

References

Bullough, Edward. 1912. "'Psychical Distance' as a Factor in Art and an Aesthetic Principle." *British Journal of Psychology* 5:87–118.

Bunge, Mario. 1963. *The Myth of Simplicity: Problems of Scientific Philosophy*. Englewood Cliffs, N.J.: Prentice-Hall.

Bush, Douglas. 1950. *Science and English Poetry: A Historical Sketch, 1590–1950*. Oxford: Oxford University Press.

Caneva, Kenneth L. 1978. "From Galvanism to Electrodynamics: The Transformation of German Physics and Its Social Context." *Historical Studies in the Physical Sciences* 9:63–159.

Cassidy, Harold G. 1962. *The Sciences and the Arts: A New Alliance*. New York: Harper.

Chandrasekhar, Subrahmanyan. 1987. *Truth and Beauty: Aesthetics and Motivations in Science*. Chicago: University of Chicago Press.

——. 1988. "A Commentary on Dirac's Views on 'The Excellence of General Relativity.' " In K. Winter, ed., *Festi–Val: Festschrift for Val Telegdi*. Amsterdam: North-Holland; pp. 49–56.

——. 1989. "The Perception of Beauty and the Pursuit of Science." *Bulletin of the American Academy of Arts and Sciences* 43, no. 3 (December): 14–29.

Channell, David F. 1991. *The Vital Machine: A Study of Technology and Organic Life*. New York: Oxford University Press.

Churchland, Paul M. 1985. "The Ontological Status of Observables: In Praise of the Superempirical Virtues." In Paul M. Churchland and Clifford A. Hooker, eds., *Images of Science: Essays on Realism and Empiricism*. Chicago: University of Chicago Press; pp. 35–47.

Cohen, G. A. 1978. *Karl Marx's Theory of History: A Defence*. Oxford: Clarendon Press.

Cohen, I. Bernard. 1980. *The Newtonian Revolution: With Illustrations of the Transformation of Scientific Ideas*. Cambridge: Cambridge University Press.

——. 1985. *Revolution in Science*. Cambridge: Harvard University Press.

Collins, Peter. 1959. *Concrete: The Vision of a New Architecture*. London: Faber and Faber.

Copernicus, Nicholas. 1543. *On the Revolutions*. Edited by Jerzy Dobrzycki and translated by Edward Rosen. In *Nicholas Copernicus: Complete Works*. 4 vols. London: Macmillan, 1972–; vol. 2 (1978).

Cossons, Neil, and Barrie Trinder. 1979. *The Iron Bridge: Symbol of the Industrial Revolution*. Bradford-on-Avon, Wiltshire: Moonraker Press.

Crombie, Alistair C. 1994. *Styles of Scientific Thinking in the European Tradition: The History of Argument and Explanation Especially in the Mathematical and Biomedical Sciences and Arts*. 3 vols. London: Duckworth.

Cronin, Helena. 1991. *The Ant and the Peacock: Altruism and Sexual Selection from Darwin to Today*. Cambridge: Cambridge University Press.

Crosnier Leconte, Marie-Laure. 1989. "La Galerie des Machines." In Musée d'Orsay (1989), pp. 164–195.

Curtin, Deane W., ed. 1982. *The Aesthetic Dimension of Science*. New York: Philosophical Library.

D'Agostino, Salvo. 1990. "Boltzmann and Hertz on the *Bild*-Conception of Physical Theory." *History of Science* 28:380–398.

References

Dalitz, R. H. 1987. "A Biographical Sketch of the Life of Professor P. A. M. Dirac, OM, FRS." In J. G. Taylor (1987), pp. 3–28.

Davies, Paul. 1992. *The Mind of God: Science and the Search for Ultimate Meaning*. London: Simon and Schuster.

Derkse, Wil. 1993. *On Simplicity and Elegance: An Essay in Intellectual History*. Delft: Eburon.

Dickie, George. 1974. *Art and the Aesthetic: An Institutional Analysis*. Ithaca: Cornell University Press.

Dirac, P. A. M. 1939. "The Relation between Mathematics and Physics." *Proceedings of the Royal Society of Edinburgh* 59:122–129.

———. 1951. "A New Classical Theory of Electrons." *Proceedings of the Royal Society of London*, ser. A, 209:291–296.

———. 1963. "The Evolution of the Physicist's Picture of Nature." *Scientific American* 208, no. 5 (May): 45–53.

———. 1977. "Recollections of an Exciting Era." In Charles Weiner, ed., *History of Twentieth Century Physics: Proceedings of the International School of Physics "Enrico Fermi," Course LVII*. New York: Academic Press; pp. 109–146.

———. 1980a. "The Excellence of Einstein's Theory of Gravitation." In Maurice Goldsmith, Alan Mackay, and James Woudhuysen, eds., *Einstein: The First Hundred Years*. Oxford: Pergamon Press; pp. 41–46.

———. 1980b. "Why We Believe in the Einstein Theory." In Bruno Gruber and Richard S. Millman, eds., *Symmetries in Science*. New York: Plenum Press; pp. 1–11.

———. 1982a. "Pretty Mathematics." *International Journal of Theoretical Physics* 21:603–605.

———. 1982b. "The Early Years of Relativity." In Holton and Elkana (1982), pp. 79–90.

Dobbs, Betty Jo Teeter. 1992. *The Janus Faces of Genius: The Role of Alchemy in Newton's Thought*. Cambridge: Cambridge University Press.

Dover, K. J. 1974. *Greek Popular Morality in the Time of Plato and Aristotle*. Oxford: Blackwell.

Duhem, Pierre. 1906. *The Aim and Structure of Physical Theory*. 2d ed., 1914. Translated by Philip P. Wiener. Princeton: Princeton University Press, 1954.

Durant, Stuart. 1994. *Palais des Machines*. London: Phaidon.

Edgerton, Samuel Y., Jr. 1991. *The Heritage of Giotto's Geometry: Art and Science on the Eve of the Scientific Revolution*. Ithaca: Cornell University Press.

Einstein, Albert. 1905. "On the Electrodynamics of Moving Bodies." Translated by Anna Beck. In *The Collected Papers of Albert Einstein*. Princeton: Princeton University Press, 1989; vol. 2, pp. 140–171.

———. 1916. "The Foundation of the General Theory of Relativity." Translated by W. Perrett and G. B. Jeffery. In *The Principle of Relativity: A Collection of Original Memoirs on the Special and General Theory of Relativity*. London: Methuen, 1923; pp. 109–164.

———. 1936. "Physics and Reality." Translated by Jean Piccard. *Journal of the Franklin Institute* 221:349–382.

———. 1949. "Autobiographical Notes." In Schilpp (1949), pp. 1–49.

Elkana, Yehuda. 1982. "The Myth of Simplicity." In Holton and Elkana (1982), pp. 205–251.

References

Elliott, Cecil D. 1992. *Technics and Architecture: The Development of Materials and Systems for Buildings*. Cambridge: MIT Press.

Engler, Gideon. 1990. "Aesthetics in Science and in Art." *British Journal of Aesthetics* 30:24–34.

———. 1994. "From Art and Science to Perception: The Role of Aesthetics." *Leonardo* 27:207–209.

Falkenburg, Brigitte. 1988. "The Unifying Role of Symmetry Principles in Particle Physics." *Ratio*, n.s. 1:113–134.

Feigl, Herbert. 1970. "Beyond Peaceful Coexistence." In Stuewer (1970), pp. 3–11.

Feuer, Lewis S. 1957. "The Principle of Simplicity." *Philosophy of Science* 24:109–122.

———. 1974. *Einstein and the Generations of Science*. 2d ed., 1982. New Brunswick, N.J.: Transaction Books.

Fleck, Ludwik. 1935. *Genesis and Development of a Scientific Fact*. Translated by Fred Bradley and Thaddeus J. Trenn. Chicago: University of Chicago Press, 1979.

Forman, Paul. 1984. "*Kausalität, Anschaulichkeit,* and *Individualität,* or How Cultural Values Prescribed the Character and the Lessons Ascribed to Quantum Mechanics." In Nico Stehr and Volker Meja, eds., *Society and Knowledge: Contemporary Perspectives in the Sociology of Knowledge*. New Brunswick, N.J.: Transaction Books; pp. 333–347.

Forster, E. M. 1927. *Aspects of the Novel*. Edited by Oliver Stallybrass. London: Edward Arnold, 1974.

Frank, Philipp. 1957. *Philosophy of Science: The Link between Science and Philosophy*. Englewood Cliffs, N.J.: Prentice-Hall.

Friebe, Wolfgang. 1983. *Buildings of the World Exhibitions*. Translated by Jenny Vowles and Paul Roper. Leipzig: Edition Leipzig, 1985.

Friedman, Alan J., and Carol C. Donley. 1985. *Einstein as Myth and Muse*. Cambridge: Cambridge University Press.

Friedman, Michael. 1974. "Explanation and Scientific Understanding." *Journal of Philosophy* 71:5–19.

———. 1983. *Foundations of Space-Time Theories: Relativistic Physics and Philosophy of Science*. Princeton: Princeton University Press.

Fry, Maxwell. 1944. *Fine Building*. London: Faber and Faber.

Galilei, Galileo. 1632. *Dialogue Concerning the Two Chief World Systems— Ptolemaic and Copernican*. Translated by Stillman Drake. Berkeley: University of California Press, 1953.

Galison, Peter L. 1979. "Minkowski's Space-Time: From Visual Thinking to the Absolute World." *Historical Studies in the Physical Sciences* 10:85–121.

Gentner, Dedre. 1983. "Structure-Mapping: A Theoretical Framework for Analogy." *Cognitive Science* 7:155–170.

Gentner, Dedre, and Michael Jeziorski. 1989. "Historical Shifts in the Use of Analogy in Science." In Barry Gholson, William R. Shadish, Jr., Robert A. Neimeyer, and Arthur C. Houts, eds., *Psychology of Science: Contributions to Metascience*. Cambridge: Cambridge University Press; pp. 296–325.

Ghiselin, Michael T. 1976. "Poetic Biology: A Defense and Manifesto." *New Literary History* 7:493–504.

Giedion, Sigfried. 1941. *Space, Time and Architecture: The Growth of a New Tradition*. 5th ed., 1967. Cambridge: Harvard University Press.

Gingerich, Owen. 1973. "The Role of Erasmus Reinhold and the Prutenic Tables in the Dissemination of Copernican Theory." In Jerzy Dobrzycki, ed., *Studia Copernicana VI*. Wroclaw: Ossolineum; pp. 43–62, 123–125.

———. 1975. " 'Crisis' versus Aesthetic in the Copernican Revolution." In Arthur Beer and K. A. Strand, eds., *Copernicus Yesterday and Today*. Vistas in Astronomy, vol. 17. Oxford: Pergamon Press; pp. 85–93.

Glick, Thomas F. 1987. "Cultural Issues in the Reception of Relativity." In Thomas F. Glick, ed., *The Comparative Reception of Relativity*. Dordrecht: Reidel; pp. 381–400.

Gloag, John. 1962. *Victorian Taste: Some Social Aspects of Architecture and Industrial Design from 1820–1900*. London: A. and C. Black.

Gloag, John, and Derek Bridgwater. 1948. *A History of Cast Iron in Architecture*. London: Allen and Unwin.

Goldman, Alan H. 1990. "Aesthetic Qualities and Aesthetic Value." *Journal of Philosophy* 87:23–37.

Gombrich, Ernst H. 1974. "The Logic of Vanity Fair: Alternatives to Historicism in the Study of Fashions, Style and Taste." In Paul A. Schilpp, ed., *The Philosophy of Karl Popper*. 2 vols. La Salle, Ill.: Open Court; vol. 2, pp. 925–957.

Grant, Edward. 1978. "Cosmology." In David C. Lindberg, ed., *Science in the Middle Ages*. Chicago: University of Chicago Press; pp. 265–302.

———. 1984. "In Defense of the Earth's Centrality and Immobility: Scholastic Reaction to Copernicanism in the Seventeenth Century." *Transactions of the American Philosophical Society* 74, pt. 4.

Gribbin, John. 1984. *In Search of Schrödinger's Cat: Quantum Physics and Reality*. New York: Bantam Books.

Gross, Alan G. 1990. *The Rhetoric of Science*. Cambridge: Harvard University Press.

Gruber, Howard E. 1978. "Darwin's 'Tree of Nature' and Other Images of Wide Scope." In Wechsler (1978), pp. 121–140.

Guedes, Pedro, ed. 1979. *The Macmillan Encyclopedia of Architecture and Technological Change*. London: Macmillan.

Haas, Arthur E. 1909. "Ästhetische und teleologische Gesichtspunkte in der antiken Physik." *Archiv für Geschichte der Philosophie* 22:80–113.

Hacking, Ian. 1992. "'Style' for Historians and Philosophers." *Studies in History and Philosophy of Science* 23:1–20.

Haldane, J. B. S. 1927. "Science and Theology as Art-Forms." In J. B. S. Haldane, *On Being the Right Size and Other Essays*. Edited by John Maynard Smith. Oxford: Oxford University Press, 1985; pp. 32–44.

Hallyn, Fernand. 1987. *The Poetic Structure of the World: Copernicus and Kepler*. Translated by Donald M. Leslie. New York: Zone Books, 1990.

Hanson, Norwood Russell. 1961. "The Copernican Disturbance and the Keplerian Revolution." *Journal of the History of Ideas* 22:169–184.

Harré, Rom. 1960. *An Introduction to the Logic of the Sciences*. 2d ed. 1983. London: Macmillan.

———. 1972. *The Philosophies of Science: An Introductory Survey*. Oxford: Oxford University Press.

References

Hatfield, Gary. 1993. "Helmholtz and Classicism: The Science of Aesthetics and the Aesthetics of Science." In David Cahan, ed., *Hermann von Helmholtz and the Foundations of Nineteenth-Century Science*. Berkeley: University of California Press; pp. 522–558.

Heilbron, John L. 1986. *The Dilemmas of an Upright Man: Max Planck as Spokesman for German Science*. Berkeley: University of California Press.

Heisenberg, Werner. 1970. "The Meaning of Beauty in the Exact Sciences." In Werner Heisenberg, *Across the Frontiers*. Translated by Peter Heath. New York: Harper and Row, 1974; pp. 166–183.

———. 1971. *Physics and Beyond: Encounters and Conversations*. Translated by Arnold J. Pomerans. New York: Harper and Row.

Helmholtz, Hermann von. 1870. "On the Origin and Significance of the Axioms of Geometry." In Robert S. Cohen and Yehuda Elkana, eds., *Hermann von Helmholtz: Epistemological Writings*. Dordrecht: Reidel, 1977; pp. 1–38.

Heninger, S. K., Jr. 1974. *Touches of Sweet Harmony: Pythagorean Cosmology and Renaissance Poetics*. San Marino, Calif.: Huntington Library.

Herrmann, Wolfgang. 1984. *Gottfried Semper: In Search of Architecture*. Cambridge: MIT Press.

Heskett, John. 1980. *Industrial Design*. London: Thames and Hudson.

Hesse, Mary B. 1954. *Science and the Human Imagination: Aspects of the History and Logic of Physical Science*. London: SCM Press.

———. 1961. *Forces and Fields: The Concept of Action at a Distance in the History of Physics*. London: Nelson.

———. 1966. *Models and Analogies in Science*. Notre Dame, Ind.: University of Notre Dame Press.

———. 1974. *The Structure of Scientific Inference*. London: Macmillan.

Hillman, Donald J. 1962. "The Measurement of Simplicity." *Philosophy of Science* 29:225–252.

Hilts, Victor L. 1975. "A Guide to Francis Galton's *English Men of Science*." *Transactions of the American Philosophical Society* 65, pt. 5.

Hoffmann, Banesh, and Helen Dukas. 1972. *Albert Einstein: Creator and Rebel*. New York: Viking Press.

Hoffmann, Roald. 1990. "Molecular Beauty." *Journal of Aesthetics and Art Criticism* 48:191–204.

Hogarth, William. 1753. *The Analysis of Beauty*. Edited by Joseph Burke. Oxford: Clarendon Press, 1955.

Holton, Gerald. 1973. *Thematic Origins of Scientific Thought: Kepler to Einstein*. Revised edition, 1988. Cambridge: Harvard University Press.

———. 1978. *The Scientific Imagination: Case Studies*. Cambridge: Cambridge University Press.

———. 1986. *The Advancement of Science and Its Burdens: The Jefferson Lecture and Other Essays*. Cambridge: Cambridge University Press.

Holton, Gerald, and Yehuda Elkana, eds. 1982. *Albert Einstein: Historical and Cultural Perspectives*. Princeton: Princeton University Press.

Honner, John. 1987. *The Description of Nature: Niels Bohr and the Philosophy of Quantum Physics*. Oxford: Clarendon Press.

Hovis, R. Corby, and Helge Kragh. 1993. "P. A. M. Dirac and the Beauty of Physics." *Scientific American* 268, no. 5 (May): 62–67.

References

215

Hoyle, Fred. 1950. *The Nature of the Universe*. Rev. ed., 1960. Harmondsworth, Middlesex: Penguin Books.

Hoyningen-Huene, Paul. 1987. "Context of Discovery and Context of Justification." *Studies in History and Philosophy of Science* 18:501–515.

Hume, David. 1739. *A Treatise of Human Nature*. Edited by L. A. Selby-Bigge and P. H. Nidditch. Oxford: Clarendon Press, 1978.

Hungerland, Isabel Creed. 1968. "Once Again, Aesthetic and Non-Aesthetic." *Journal of Aesthetics and Art Criticism* 26:285–295.

Huntley, H. E. 1970. *The Divine Proportion: A Study in Mathematical Beauty*. New York: Dover.

Hutcheson, Francis. 1725. *An Inquiry Concerning Beauty, Order, Harmony, Design*. Edited by Peter Kivy. The Hague: Martinus Nijhoff, 1973.

Hutchison, Keith. 1982. "What Happened to Occult Qualities in the Scientific Revolution?" *Isis* 73:233–253.

——. 1987. "Towards a Political Iconology of the Copernican Revolution." In Patrick Curry, ed., *Astrology, Science and Society: Historical Essays*. Woodbridge, Suffolk: Boydell; pp. 95–141.

Huxley, Thomas Henry. 1894. "Biogenesis and Abiogenesis." In Thomas Henry Huxley, *Collected Essays*. 9 vols. London: Macmillan; vol. 8, pp. 229–271.

Jacquette, Dale. 1990. "Aesthetics and Natural Law in Newton's Methodology." *Journal of the History of Ideas* 51:659–666.

Jammer, Max. 1966. *The Conceptual Development of Quantum Mechanics*. New York: McGraw-Hill.

——. 1974. *The Philosophy of Quantum Mechanics: The Interpretations of Quantum Mechanics in Historical Perspective*. New York: Wiley.

Jardine, Nicholas. 1982. "The Significance of the Copernican Orbs." *Journal for the History of Astronomy* 13:168–194.

——. 1991. *The Scenes of Inquiry: On the Reality of Questions in the Sciences*. Oxford: Clarendon Press.

Kaiser, David. 1994. "Bringing the Human Actors Back on Stage: The Personal Context of the Einstein–Bohr Debate." *British Journal for the History of Science* 27:129–152.

Kargon, Robert. 1969. "Model and Analogy in Victorian Science: Maxwell's Critique of the French Physicists." *Journal of the History of Ideas* 30:423–436.

Kargon, Robert, and Peter Achinstein, eds. 1987. *Kelvin's Baltimore Lectures and Modern Theoretical Physics: Historical and Philosophical Perspectives*. Cambridge: MIT Press.

Keller, Alex. 1983. *The Infancy of Atomic Physics: Hercules in His Cradle*. Oxford: Clarendon Press.

Kemeny, John G. 1953. "The Use of Simplicity in Induction." *Philosophical Review* 62:391–408.

Kemp, Martin. 1990. *The Science of Art: Optical Themes in Western Art from Brunelleschi to Seurat*. New Haven: Yale University Press.

Kepler, Johannes. 1609. *New Astronomy*. Translated by William H. Donahue. Cambridge: Cambridge University Press, 1993.

——. 1627. *Tabulae Rudolphinae*. Edited by Franz Hammer. In Max Caspar and

References

Franz Hammer, eds., *Johannes Kepler: Gesammelte Werke*. 22 vols. Munich: C. H. Beck, 1937–; vol. 10 (1969).

King, Jerry P. 1992. *The Art of Mathematics*. New York: Plenum Press.

Kippenhahn, Rudolph. 1984. *Light from the Depths of Time*. Translated by Storm Dunlop. Berlin: Springer-Verlag, 1987.

Kitcher, Philip. 1989. "Explanatory Unification and the Causal Structure of the World." In Philip Kitcher and Wesley C. Salmon, eds., *Scientific Explanation*. Minnesota Studies in the Philosophy of Science, vol. 13. Minneapolis: University of Minnesota Press; pp. 410–506.

Kivy, Peter. 1976. *The Seventh Sense: A Study of Francis Hutcheson's Aesthetics and Its Influence in Eighteenth-Century Britain*. New York: Burt Franklin.

———. 1991. "Science and Aesthetic Appreciation." In Peter A. French, Theodore E. Uehling, Jr., and Howard K. Wettstein, eds., *Philosophy and the Arts*. Midwest Studies in Philosophy, vol. 16. Notre Dame, Ind.: University of Notre Dame Press; pp. 180–195.

Klein, Martin J. 1972. "Mechanical Explanation at the End of the Nineteenth Century." *Centaurus* 17:58–82.

Koestler, Arthur. 1959. *The Sleepwalkers: A History of Man's Changing Vision of the Universe*. London: Hutchinson.

Kordig, Carl R. 1971. *The Justification of Scientific Change*. Dordrecht: Reidel.

Koyré, Alexandre. 1939. *Galileo Studies*. Translated by John Mepham. Hassocks, Sussex: Harvester Press, 1978.

———. 1955. "Attitude esthétique et pensée scientifique." *Critique: Revue générale des publications françaises et étrangères* 11:835–847.

Kragh, Helge. 1990. *Dirac: A Scientific Biography*. Cambridge: Cambridge University Press.

Krisch, A. D. 1987. "An Experimenter's View of P. A. M. Dirac." In Kursunoglu and Wigner (1987), pp. 46–52.

Kuhn, Thomas S. 1957. *The Copernican Revolution: Planetary Astronomy in the Development of Western Thought*. Cambridge: Harvard University Press.

———. 1962. *The Structure of Scientific Revolutions*. 2d ed., 1970. Chicago: University of Chicago Press.

———. 1970. "Reflections on My Critics." In Lakatos and Musgrave (1970), pp. 231–278.

———. 1977. *The Essential Tension: Selected Studies in Scientific Tradition and Change*. Chicago: University of Chicago Press.

Kursunoglu, Behram N., and Eugene P. Wigner, eds. 1987. *Reminiscences about a Great Physicist: Paul Adrien Maurice Dirac*. Cambridge: Cambridge University Press.

Lagrange, Joseph Louis. 1788. *Mécanique analytique*, vol. 1. 2d ed., 1811. Reprinted in J. A. Serret, ed., *Oeuvres de Lagrange*. 14 vols. Paris: Gauthier-Villars, 1867–1892; vol. 11 (1888).

Lakatos, Imre. 1970. "Falsification and the Methodology of Scientific Research Programmes." In Lakatos and Musgrave (1970), pp. 91–196.

———. 1971. "History of Science and Its Rational Reconstruction." In Roger C. Buck and Robert S. Cohen, eds., *PSA 1970: Proceedings of the 1970 Biennial Meeting, Philosophy of Science Association*. Dordrecht: Reidel; pp. 91–136.

References

Lakatos, Imre, and Alan Musgrave, eds. 1970. *Criticism and the Growth of Knowledge*. Cambridge: Cambridge University Press.

Lamouche, André. 1955. *Le Principe de simplicité dans les mathématiques et dans les sciences physiques*. Paris: Gauthier-Villars.

Laplace, Pierre-Simon de. 1813. *Exposition du système du monde*. 4th ed. 2 vols. Paris: Courcier.

Latour, Bruno. 1984. *The Pasteurization of France*. Translated by Alan Sheridan and John Law. Cambridge: Harvard University Press, 1988.

——. 1987. *Science in Action: How to Follow Scientists and Engineers Through Society*. Milton Keynes, Buckinghamshire: Open University Press.

Latour, Bruno, and Steve Woolgar. 1979. *Laboratory Life: The Construction of Scientific Facts*. 2d ed., 1986. Princeton: Princeton University Press.

Laudan, Larry. 1977. *Progress and Its Problems: Towards a Theory of Scientific Growth*. Berkeley: University of California Press.

——. 1981. *Science and Hypothesis: Historical Essays on Scientific Methodology*. Dordrecht: Reidel.

——. 1984. *Science and Values: The Aims of Science and Their Role in Scientific Debate*. Berkeley: University of California Press.

Leary, David E. 1990a. "Psyche's Muse: The Role of Metaphor in the History of Psychology." In Leary (1990b), pp. 1–78.

——, ed. 1990b. *Metaphors in the History of Psychology*. Cambridge: Cambridge University Press.

Le Lionnais, François. 1948. "Beauty in Mathematics." In François Le Lionnais, ed., *Great Currents of Mathematical Thought*. 2d ed., 1962. Translated by R. A. Hall et al. 2 vols. New York: Dover, 1971; vol. 2, pp. 121–158.

Levine, Neil. 1977. "The Romantic Idea of Architectural Legibility: Henri Labrouste and the Néo-Grec." In Arthur Drexler, ed., *The Architecture of the Ecole des Beaux-Arts*. London: Secker and Warburg; pp. 325–416.

——. 1982. "The Book and the Building: Hugo's Theory of Architecture and Labrouste's Bibliothèque Ste-Geneviève." In Robin Middleton, ed., *The Beaux-Arts and Nineteenth-Century French Architecture*. London: Thames and Hudson; pp. 138–173.

Levins, Richard, and Richard Lewontin. 1985. *The Dialectical Biologist*. Cambridge: Harvard University Press.

Lewis, David. 1973. *Counterfactuals*. Oxford: Blackwell.

Li, Ming, and Paul M. B. Vitányi. 1992. "Inductive Reasoning and Kolmogorov Complexity." *Journal of Computer and System Sciences* 44:343–384.

Lindeboom, G. A. 1968. *Herman Boerhaave: The Man and His Work*. London: Methuen.

Lipscomb, William N., Jr. 1982. "Aesthetic Aspects of Science." In Curtin (1982), pp. 1–24.

Lodge, Oliver. 1883. "The Ether and Its Functions," parts 1 and 2. *Nature* 27:304–306, 328–330.

Lorentz, H. A. 1920. *The Einstein Theory of Relativity: A Concise Statement*. New York: Brentano's.

Lovejoy, Arthur O. 1936. *The Great Chain of Being: A Study of the History of an Idea*. 2d ed., 1964. Cambridge: Harvard University Press.

References

Loyrette, Henri. 1985. *Gustave Eiffel*. Translated by Rachel Gomme and Susan Gomme. New York: Rizzoli.

——. 1989. "Images de la Tour Eiffel (1884–1914)." In Musée d'Orsay (1989), pp. 196–219.

Lumsden, Charles J. 1991. "Aesthetics." In Mary Maxwell, ed., *The Sociobiological Imagination*. Albany: State University of New York Press; pp. 253–268.

Lynch, Michael, and Samuel Y. Edgerton, Jr. 1988. "Aesthetics and Digital Image Processing: Representational Craft in Contemporary Astronomy." In Gordon Fyfe and John Law, eds., *Picturing Power: Visual Depiction and Social Relations*. London: Routledge; pp. 184–220.

Mach, Ernst. 1883. *The Science of Mechanics: A Critical and Historical Account of Its Development*. Translated by Thomas J. McCormack. 6th ed., 1960. La Salle, Ill.: Open Court.

Machan, Tibor R. 1977. "Kuhn, Paradigm Choice and the Arbitrariness of Aesthetic Criteria in Science." *Theory and Decision* 8:361–362.

Mackie, John L. 1977. *Ethics: Inventing Right and Wrong*. Harmondsworth, Middlesex: Penguin Books.

Mamchur, Elena. 1987. "The Heuristic Role of Aesthetics in Science." *International Studies in the Philosophy of Science* 1:209–222.

Margenau, Henry. 1950. *The Nature of Physical Reality: A Philosophy of Modern Physics*. New York: McGraw-Hill.

Mark, Robert, and David P. Billington. 1989. "Structural Imperative and the Origin of New Form." *Technology and Culture* 30:300–329.

Martin, James E. 1989. "Aesthetic Constraints on Theory Selection: A Critique of Laudan." *British Journal for the Philosophy of Science* 40:357–364.

Maxwell, James Clerk. 1873. *A Treatise on Electricity and Magnetism*. 3d ed., 1892. 2 vols. Oxford: Clarendon Press.

McAllister, James W. 1989. "Truth and Beauty in Scientific Reason." *Synthese* 78:25–51.

——. 1990. "Dirac and the Aesthetic Evaluation of Theories." *Methodology and Science* 23:87–102.

——. 1991a. "Scientists' Aesthetic Judgements." *British Journal of Aesthetics* 31:332–341.

——. 1991b. "The Simplicity of Theories: Its Degree and Form." *Journal for General Philosophy of Science* 22:1–14.

——. 1993. "Scientific Realism and the Criteria for Theory-Choice." *Erkenntnis* 38:203–222.

——. 1995. "The Formation of Styles: Science and the Applied Arts." In Caroline A. van Eck, James W. McAllister, and Renée van de Vall, eds., *The Question of Style in Philosophy and the Arts*. Cambridge: Cambridge University Press; pp. 157–176.

——. 1996. "Scientists' Aesthetic Preferences among Theories: Conservative Factors in Revolutionary Crises." In Tauber (1996), pp. 169–187.

McDowell, John. 1983. "Aesthetic Value, Objectivity, and the Fabric of the World." In Eva Schaper, ed., *Pleasure, Preference and Value: Studies in Philosophical Aesthetics*. Cambridge: Cambridge University Press; pp. 1–16.

McKean, John. 1994. *Crystal Palace*. London: Phaidon.

References

McReynolds, Paul. 1990. "Motives and Metaphors: A Study in Scientific Creativity." In Leary (1990b), pp. 133–172.

Mehra, Jagdish. 1972. "'The Golden Age of Theoretical Physics': P. A. M. Dirac's Scientific Work from 1924 to 1933." In Salam and Wigner (1972), pp. 17–59.

Mehra, Jagdish, and Helmut Rechenberg. 1982–1987. *The Historical Development of Quantum Theory*. 5 vols. New York: Springer-Verlag.

Mellor, D. H. 1968. "Models and Analogies in Science: Duhem *versus* Campbell?" *Isis* 59:282–290.

——. 1988. "The Warrant of Induction." Reprinted in D. H. Mellor, *Matters of Metaphysics*. Cambridge: Cambridge University Press, 1991; pp. 254–268.

Michelis, P. A. 1963. *Esthétique de l'architecture du béton armé*. Paris: Dunod.

Miller, Arthur I. 1981. *Albert Einstein's Special Theory of Relativity: Emergence (1905) and Early Interpretation (1905–1911)*. Reading, Mass.: Addison-Wesley.

——. 1984. *Imagery in Scientific Thought: Creating 20th-Century Physics*. Boston: Birkhäuser.

Mittelstrass, Jürgen. 1972. "Methodological Elements of Keplerian Astronomy." *Studies in History and Philosophy of Science* 3:203–232.

Moesgaard, Kristian P. 1972. "Copernican Influence on Tycho Brahe." In Jerzy Dobrzycki, ed., *The Reception of Copernicus' Heliocentric Theory*. Dordrecht: Reidel; pp. 31–55.

Moore, Walter. 1989. *Schrödinger: Life and Thought*. Cambridge: Cambridge University Press.

Mott, Nevill. 1986. *A Life in Science*. London: Taylor and Francis.

Murdoch, Dugald. 1987. *Niels Bohr's Philosophy of Physics*. Cambridge: Cambridge University Press.

Musée d'Orsay. 1989. *1889: La Tour Eiffel et l'Exposition Universelle*. Paris: Éditions de la Réunion des Musées Nationaux.

Nagel, Ernest. 1961. *The Structure of Science: Problems in the Logic of Scientific Explanation*. London: Routledge and Kegan Paul.

Nature. 1927. "News and Views", unsigned, in *Nature* 119:467–471.

Nersessian, Nancy J. 1988. "Reasoning from Imagery and Analogy in Scientific Concept Formation." In Arthur Fine and Jarrett Leplin, eds., *PSA 1988: Proceedings of the 1988 Biennial Meeting of the Philosophy of Science Association*. 2 vols. East Lansing, Mich.: Philosophy of Science Association; vol. 1, pp. 41–47.

Neugebauer, Otto. 1952. *The Exact Sciences in Antiquity*. 2d ed., 1969. New York: Dover.

——. 1968. "On the Planetary System of Copernicus." In Arthur Beer, ed., *Philosophy, Dynamics, Astrometry, Astro-Archaeology, Correlations, Astrophysics, History, Instrumentation, Cosmogony*. Vistas in Astronomy, vol. 10. Oxford: Pergamon Press; pp. 89–103.

Newton-Smith, W. H. 1981. *The Rationality of Science*. London: Routledge and Kegan Paul.

Neyman, Jerzy. 1974. "Nicholas Copernicus (Mikolaj Kopernik): An Intellectual Revolutionary." In Jerzy Neyman, ed., *The Heritage of Copernicus: Theories "Pleasing to the Mind."* Cambridge: MIT Press; pp. 1–22.

References

Nicolson, Marjorie Hope. 1946. *Newton Demands the Muse: Newton's "Opticks" and the Eighteenth-Century Poets*. Princeton: Princeton University Press.

——. 1950. *The Breaking of the Circle: Studies in the Effect of the "New Science" upon Seventeeth-Century Poetry*. Rev. ed., 1960. New York: Columbia University Press.

——. 1956. *Science and Imagination*. Ithaca: Cornell University Press.

Nisbet, Robert. 1976. *Sociology as an Art Form*. New York: Oxford University Press.

Osborne, Harold. 1964. "Notes on the Aesthetics of Chess and the Concept of Intellectual Beauty." *British Journal of Aesthetics* 4:160–163.

——. 1984. "Mathematical Beauty and Physical Science." *British Journal of Aesthetics* 24:291–300.

——. 1986a. "Interpretation in Science and in Art." *British Journal of Aesthetics* 26:3–15.

——. 1986b. "Symmetry as an Aesthetic Factor." *Computers and Mathematics with Applications*, ser. B, 12:77–82.

Pais, Abraham. 1982. *"Subtle Is the Lord . . .": The Science and the Life of Albert Einstein*. Oxford: Clarendon Press.

——. 1991. *Niels Bohr's Times, in Physics, Philosophy, and Polity*. Oxford: Clarendon Press.

Palter, Robert. 1970. "An Approach to the History of Early Astronomy." *Studies in History and Philosophy of Science* 1:93–133.

Panofsky, Erwin. 1954. *Galileo as a Critic of the Arts*. The Hague: Martinus Nijhoff.

——. 1956. "Galileo as a Critic of the Arts: Aesthetic Attitude and Scientific Thought." *Isis* 47:3–15.

Papert, Seymour A. 1978. "The Mathematical Unconscious." In Wechsler (1978), pp. 105–119.

Pawley, Martin. 1990. *Theory and Design in the Second Machine Age*. Oxford: Blackwell.

Penrose, Roger. 1974. "The Rôle of Aesthetics in Pure and Applied Mathematical Research." *Bulletin of the Institute of Mathematics and Its Applications* 10:266–271.

Pera, Marcello. 1981. "Copernico e il realismo scientifico." *Filosofia* 22:151–174.

Pevsner, Nikolaus. 1937. *An Enquiry into Industrial Art in England*. Cambridge: Cambridge University Press.

——. 1960. *Pioneers of Modern Design: From William Morris to Walter Gropius*. Rev. ed., 1974. Harmondsworth, Middlesex: Penguin Books.

——. 1968. *The Sources of Modern Architecture and Design*. London: Thames and Hudson.

Pickering, Andrew, ed. 1992. *Science as Practice and Culture*. Chicago: University of Chicago Press.

Pickvance, Simon. 1986. "'Life' in a Biology Lab." In Les Levidow, ed., *Radical Science Essays*. London: Free Association Books; pp. 140–153.

Planck, Max. 1922. *The Origin and Development of the Quantum Theory*. Translated by H. T. Clarke and L. Silberstein. Oxford: Clarendon Press.

——. 1948. "A Scientific Autobiography." In Max Planck, *Scientific Autobiogra-*

References

221

phy and Other Papers. Translated by Frank Gaynor. London: Williams and Norgate, 1950; pp. 13–51.

Pocock, Stuart J. 1983. *Clinical Trials: A Practical Approach*. Chichester, West Sussex: Wiley.

Poincaré, Henri. 1908. *Science and Method*. Translated by F. Maitland. London: Nelson, 1914.

Polanyi, Michael. 1958. *Personal Knowledge: Towards a Post-Critical Philosophy*. London: Routledge and Kegan Paul.

Pollitt, J. J. 1974. *The Ancient View of Greek Art: Criticism, History, and Terminology*. New Haven: Yale University Press.

Popper, Karl R. 1959. *The Logic of Scientific Discovery*. Rev. ed., 1980. London: Hutchinson.

——. 1972. *Objective Knowledge: An Evolutionary Approach*. Oxford: Clarendon Press.

Pribram, Karl H. 1990. "From Metaphors to Models: The Use of Analogy in Neuropsychology." In Leary (1990b), pp. 79–103.

Price, Derek J. de S. 1959. "Contra-Copernicus: A Critical Re-Estimation of the Mathematical Planetary Theory of Ptolemy, Copernicus, and Kepler." In Marshall Clagett, ed., *Critical Problems in the History of Science*. Madison: University of Wisconsin Press; pp. 197–218.

Price, Kingsley. 1977. "The Truth about Psychical Distance." *Journal of Aesthetics and Art Criticism* 35:411–423.

Priest, Graham. 1976. "Gruesome Simplicity." *Philosophy of Science* 43:432–437.

Quine, Willard Van Orman. 1953. *From a Logical Point of View: Nine Logico-Philosophical Essays*. 2d ed., 1961. Cambridge: Harvard University Press.

Randall, John H., Jr. 1960. *Aristotle*. New York: Columbia University Press.

Ray, Christopher. 1987. *The Evolution of Relativity*. Bristol: Adam Hilger.

Reichenbach, Hans. 1927. *From Copernicus to Einstein*. Translated by Ralph B. Winn. New York: Dover, 1980.

——. 1938. *Experience and Prediction: An Analysis of the Foundations and the Structure of Knowledge*. Chicago: University of Chicago Press.

Rescher, Nicholas. 1977. *Methodological Pragmatism: A Systems-Theoretic Approach to the Theory of Knowledge*. Oxford: Blackwell.

——. 1987. *Scientific Realism: A Critical Reappraisal*. Dordrecht: Reidel.

——, ed. 1990. *Aesthetic Factors in Natural Science*. Lanham, Md.: University Press of America.

Rheticus, Georg Joachim. 1540. *Narratio prima*. Translated by Edward Rosen. In Edward Rosen, ed., *Three Copernican Treatises*. 2d ed., 1959. New York: Dover, 1939; pp. 107–196.

Rigden, John S. 1986. "Quantum States and Precession: The Two Discoveries of NMR." *Reviews of Modern Physics* 58:433–448.

Rohrlich, Fritz. 1987. *From Paradox to Reality: Our Basic Concepts of the Physical World*. Cambridge: Cambridge University Press.

Rolston, Holmes, III. 1995. "Does Aesthetic Appreciation of Landscapes Need to Be Science-Based?" *British Journal of Aesthetics* 35:374–386.

Root-Bernstein, Robert S. 1984. "On Paradigms and Revolutions in Science and Art: The Challenge of Interpretation." *Art Journal* 44:109–118.

References

——. 1985. "Visual Thinking: The Art of Imagining Reality." *Transactions of the American Philosophical Society* 75, no. 6: 50–67.

——. 1987. "Harmony and Beauty in Medical Research." *Journal of Molecular and Cellular Cardiology* 19:1043–1051.

Rose, Paul L. 1975. "Universal Harmony in Regiomontanus and Copernicus." In Suzanne Delorme, ed., *Avant, avec, après Copernic: La Représentation de l'univers et ses conséquences épistémologiques*. Paris: Blanchard; pp. 153–158.

Rosen, Joe. 1975. *Symmetry Discovered: Concepts and Applications in Nature and Science*. Cambridge: Cambridge University Press.

Rosenfeld, Léon. 1967. "Niels Bohr in the Thirties: Consolidation and Extension of the Conception of Complementarity." In Stefan Rozental, ed., *Niels Bohr: His Life and Work as Seen by His Friends and Colleagues*. Amsterdam: North-Holland; pp. 114–136.

Rosenkrantz, R. D. 1976. "Simplicity." In William L. Harper and Clifford A. Hooker, eds., *Foundations of Probability Theory, Statistical Inference, and Statistical Theories of Science*. 3 vols. Dordrecht: Reidel; vol. 1, pp. 167–196.

Rosenthal-Schneider, Ilse. 1980. "Reminiscences of Einstein." In Harry Woolf, ed., *Some Strangeness in the Proportion: A Centennial Symposium to Celebrate the Achievements of Albert Einstein*. Reading, Mass.: Addison-Wesley; pp. 521–523.

Ruskin, John. 1849. *The Seven Lamps of Architecture*. 2d ed., 1880. Reprint. London: George Allen and Sons, 1907.

Russell, Bertrand. 1940. "The Philosophy of Santayana." In Paul A. Schilpp, ed., *The Philosophy of George Santayana*. Evanston, Ill.: Northwestern University Press; pp. 453–474.

Russell, John L. 1964. "Kepler's Laws of Planetary Motion: 1609–1666." *British Journal for the History of Science* 2:1–24.

Salam, Abdus, and Eugene P. Wigner, eds. 1972. *Aspects of Quantum Theory*. Cambridge: Cambridge University Press.

Salmon, Wesley C. 1961. "Comments on Barker's 'The Role of Simplicity in Explanation.' " In Herbert Feigl and Grover Maxwell, eds., *Current Issues in the Philosophy of Science*. New York: Holt, Rinehart and Winston; pp. 274–276.

Scheffler, Israel. 1967. *Science and Subjectivity*. Indianapolis: Bobbs-Merrill.

Schilpp, Paul A., ed. 1949. *Albert Einstein: Philosopher-Scientist*. La Salle, Ill.: Open Court.

Schrödinger, Erwin. 1958. *Mind and Matter*. Cambridge: Cambridge University Press.

Schweber, Silvan S. 1994. *QED and the Men Who Made It: Dyson, Feynman, Schwinger, and Tomonaga*. Princeton: Princeton University Press.

Sellar, Walter C., and Robert J. Yeatman. 1930. *1066 and All That: A Memorable History of England*. London: Methuen.

Shaftesbury, Anthony Ashley Cooper, Third Earl of. 1711. *Characteristics of Men, Manners, Opinions, Times etc.* Edited by John M. Robertson. 2 vols. London: Grant Richards, 1900.

Shanmugadhasan, S. 1987. "Dirac as Research Supervisor and Other Remembrances." In J. G. Taylor (1987), pp. 48–57.

Shea, William R. 1985. "Panofsky Revisited: Galileo as a Critic of the Arts." In

References

Andrew Morrogh, Fiorella Superbi Gioffredi, Piero Morselli, and Eve Borsook, eds., *Renaissance Studies in Honor of Craig Hugh Smyth.* 2 vols. Florence: Giunti Barbèra; vol. 1, pp. 481–492.

Shelton, Jim. 1988. "The Role of Observation and Simplicity in Einstein's Epistemology." *Studies in History and Philosophy of Science* 19:103–118.

Shepard, Roger N. 1978. "The Mental Image." *American Psychologist* 33:125–137.

Sibley, Frank. 1959. "Aesthetic Concepts." *Philosophical Review* 68:421–450.

Simmons, Jack. 1968. *St. Pancras Station.* London: Allen and Unwin.

Simonton, Dean K. 1988. *Scientific Genius: A Psychology of Science.* Cambridge: Cambridge University Press.

Skempton, A. W., and H. R. Johnson. 1962. "The First Iron Frames." *Architectural Review* 131:175–186.

Smith, Adam. 1776. *An Inquiry into the Nature and Causes of the Wealth of Nations.* Edited by R. H. Campbell, A. S. Skinner, and W. B. Todd. Oxford: Clarendon Press, 1976.

Sober, Elliott. 1975. *Simplicity.* Oxford: Clarendon Press.

——. 1984. *The Nature of Selection: Evolutionary Theory in Philosophical Focus.* Cambridge: MIT Press.

——. 1988. *Reconstructing the Past: Parsimony, Evolution, and Inference.* Cambridge: MIT Press.

Sparke, Penny. 1986a. *An Introduction to Design and Culture in the Twentieth Century.* London: Allen and Unwin.

——. 1986b. *Furniture.* London: Bell and Hyman.

Stachel, John. 1986. "Einstein and the Quantum: Fifty Years of Struggle." In Robert G. Colodny, ed., *From Quarks to Quasars: Philosophical Problems of Modern Physics.* Pittsburgh: University of Pittsburgh Press; pp. 349–385.

Stafford, Barbara M. 1984. *Voyage into Substance: Art, Science, Nature, and the Illustrated Travel Account, 1760–1840.* Cambridge: MIT Press.

Stolnitz, Jerome. 1960. *Aesthetics and Philosophy of Art Criticism: A Critical Introduction.* Boston: Houghton Mifflin.

——. 1961a. "On the Significance of Lord Shaftesbury in Modern Aesthetic Theory." *Philosophical Quarterly* 11:97–113.

——. 1961b. "On the Origins of 'Aesthetic Disinterestedness.' " *Journal of Aesthetics and Art Criticism* 20:131–143.

Strike, James. 1991. *Construction into Design: The Influence of New Methods of Construction on Architectural Design, 1690–1990.* Oxford: Butterworth Architecture.

Stuewer, Roger H., ed. 1970. *Historical and Philosophical Perspectives of Science.* Minnesota Studies in the Philosophy of Science, vol. 5. Minneapolis: University of Minnesota Press.

Sullivan, J. W. N. 1919. "The Justification of the Scientific Method." *The Athenaeum*, no. 4644 (2 May): 274–275.

Swenson, Loyd S., Jr. 1972. *The Ethereal Aether: A History of the Michelson-Morley-Miller Aether-Drift Experiments, 1880–1930.* Austin: University of Texas Press.

Swerdlow, Noel M. 1973. "The Derivation and First Draft of Copernicus's

References

Planetary Theory: A Translation of the *Commentariolus* with Commentary." *Proceedings of the American Philosophical Society* 117:423–512.

Swerdlow, Noel M., and Otto Neugebauer. 1984. *Mathematical Astronomy in Copernicus's De Revolutionibus.* 2 vols. New York: Springer-Verlag.

Tauber, Alfred I., ed. 1996. *The Elusive Synthesis: Aesthetics and Science.* Dordrecht: Kluwer.

Taylor, A. M. 1966. *Imagination and the Growth of Science.* London: Murray.

Taylor, J. G., ed. 1987. *Tributes to Paul Dirac.* Bristol: Adam Hilger.

Thagard, Paul. 1988. *Computational Philosophy of Science.* Cambridge: MIT Press.

Thomson, George. 1961. *The Inspiration of Science.* Oxford: Oxford University Press.

Thomson, Herbert F. 1965. "Adam Smith's Philosophy of Science." *Quarterly Journal of Economics* 79:212–233.

Thomson, William, Lord Kelvin. 1884. *Notes of Lectures on Molecular Dynamics and the Wave Theory of Light.* Reprinted in Kargon and Achinstein (1987), pp. 7–263.

———. 1901. "Nineteenth-Century Clouds over the Dynamical Theory of Heat and Light." Reprinted in Lord Kelvin, *Baltimore Lectures on Molecular Dynamics and the Wave Theory of Light.* Cambridge: Cambridge University Press, 1904; pp. 486–527.

Tonietti, Tito. 1985. Letter to the Editor. *Mathematical Intelligencer* 7, no. 4: 6–8.

Tsilikis, John D. 1959. "Simplicity and Elegance in Theoretical Physics." *American Scientist* 47:87–96.

van Fraassen, Bas C. 1980. *The Scientific Image.* Oxford: Clarendon Press.

———. 1989. *Laws and Symmetry.* Oxford: Clarendon Press.

Van Vleck, John H. 1972. "Travels with Dirac in the Rockies." In Salam and Wigner (1972), pp. 7–16.

Van Zanten, David. 1987. *Designing Paris: The Architecture of Duban, Labrouste, Duc, and Vaudoyer.* Cambridge: MIT Press.

Veltman, Kim H. 1986. *Linear Perspective and the Visual Dimensions of Science and Art.* Munich: Deutscher Kunstverlag.

Viollet-le-Duc, Eugène Emmanuel. 1863–1872. *Entretiens sur l'architecture.* 2 vols. Facsimile edition. Paris: Pierre Mardaga, 1977.

Walsh, Dorothy. 1979. "Occam's Razor: A Principle of Intellectual Elegance." *American Philosophical Quarterly* 16:241–244.

Watkins, John. 1958. "Confirmable and Influential Metaphysics." *Mind* 67:344–365.

———. 1984. *Science and Scepticism.* Princeton: Princeton University Press.

Watson, James D. 1968. *The Double Helix: A Personal Account of the Discovery of the Structure of DNA.* Critical edition, edited by Gunther S. Stent. London: Weidenfeld and Nicolson, 1981.

Webster, Charles. 1982. *From Paracelsus to Newton: Magic and the Making of Modern Science.* Cambridge: Cambridge University Press.

Wechsler, Judith, ed. 1978. *On Aesthetics in Science.* Cambridge: MIT Press.

Weinberg, Steven. 1987. "Towards the Final Laws of Physics." In Richard P. Feynman and Steven Weinberg, *Elementary Particles and the Laws of Physics: The 1986 Dirac Memorial Lectures.* Cambridge: Cambridge University Press; pp. 61–110.

References

———. 1993. *Dreams of a Final Theory*. London: Hutchinson.

Wessels, Linda. 1983. "Erwin Schrödinger and the Descriptive Tradition." In Rutherford Aris, H. Ted Davis, and Roger H. Stuewer, eds., *Springs of Scientific Creativity: Essays on Founders of Modern Science*. Minneapolis: University of Minnesota Press; pp. 254–278.

Wessely, Anna. 1991. "Transposing 'Style' from the History of Art to the History of Science." *Science in Context* 4:265–278.

Westman, Robert S. 1975. "The Melanchthon Circle, Rheticus, and the Wittenberg Interpretation of the Copernican Theory." *Isis* 66:165–193.

———. 1990. "Proof, Poetics, and Patronage: Copernicus's Preface to *De Revolutionibus*." In David C. Lindberg and Robert S. Westman, eds., *Reappraisals of the Scientific Revolution*. Cambridge: Cambridge University Press; pp. 167–205.

Weyl, Hermann. 1952. *Symmetry*. Princeton: Princeton University Press.

Wheaton, Bruce R. 1983. *The Tiger and the Shark: Empirical Roots of Wave–Particle Dualism*. Cambridge: Cambridge University Press.

Wheeler, John A. 1983. "Law without Law." In John A. Wheeler and W. H. Zurek, eds., *Quantum Theory and Measurement*. Princeton: Princeton University Press; pp. 182–213.

Whiteside, D. T. 1974. "Keplerian Planetary Eggs, Laid and Unlaid, 1600–1605." *Journal for the History of Astronomy* 5:1–21.

Whitrow, Gerald J. 1967. *Einstein: The Man and His Achievement*. London: British Broadcasting Corporation.

Wiener, Norbert. 1948. *Cybernetics, or Control and Communication in the Animal and the Machine*. 2d ed., 1961. Cambridge: MIT Press.

Williams, George C. 1966. *Adaptation and Natural Selection: A Critique of Some Current Evolutionary Thought*. Princeton: Princeton University Press.

Williamson, Robert B. 1977. "Logical Economy in Einstein's 'On the Electrodynamics of Moving Bodies.'" *Studies in History and Philosophy of Science* 8:49–60.

Wilson, Curtis. 1968. "Kepler's Derivation of the Elliptical Path." *Isis* 59:4–25.

Wilson, Edward O. 1978. *On Human Nature*. Cambridge: Harvard University Press.

Wollheim, Richard. 1968. *Art and Its Objects*. 2d ed., 1980. Cambridge: Cambridge University Press.

Yang, Chen Ning. 1961. *Elementary Particles: A Short History of Some Discoveries in Atomic Physics*. Princeton: Princeton University Press.

Zee, Anthony. 1986. *Fearful Symmetry: The Search for Beauty in Modern Physics*. New York: Macmillan.

Zemach, Eddy M. 1986. "Truth and Beauty." *Philosophical Forum* 18:21–39.

References

Index

abstractness, 49–54, 189–195
actor-network theory, 26–28
analogy. *See* models, invocation of
Apollonius of Perga, 168
aptness and beauty, 37–38
architecture, 142–162
Aristotelianism:
  classification of sciences, 165
  cosmology, 20, 168–171
  natural philosophy, 95, 174
Arkwright, Richard, 143
art:
  applied, 142–162
  science and, 132–133, 159–162
  scientific theories and, 14, 20–21, 25
astronomy. *See* Copernican theory; Keplerian theory; Ptolemaic theory
atomic theory, 45, 65, 189, 196
authenticity, aesthetic, 145, 150–155, 159, 162

Bachelard, Gaston, 126
Baird, John, 148
Barker, Stephen F., 106
Barlow, William H., 145
Bayesian theory, 107
beauty:
  and aesthetic properties, 32–34
  as aesthetic value, 30–33
  and aptness, 37–38
  intellectual, 17–18
  mathematical, 15–16, 21–23, 53, 59, 190–192
  natural, 19–21, 23
  and truth, 67–69, 90–104

Bergmann, Peter G., 188
Bernstein, Jeremy, 84
biological sciences, 47–48, 59–60, 91, 117–118
Boerhaave, Herman, 111
Bohr, Niels:
  atomic theory, 65, 189, 196
  and Einstein, 134–135, 198, 200
  as positivist, 132, 200
  on visualization, 192–194
Boltzmann, Ludwig, 87–88, 125
Boscovich, Roger, 111
Boulton, Matthew, 143
Boyle, Robert, 56
Bragg, William H., 194
Brahe, Tycho, 173, 178
Braithwaite, Richard B., 100
Broglie, Louis de, 43, 190
Brunel, Isambard Kingdom, 143
Buchdahl, Gerd, 107
Bullough, Edward, 63–64

caloric theory, 45
Campbell, Norman R., 51–52
canons, aesthetic:
  composition of, 34–35
  conservatism of, 81–85, 128–133
  formation of, 78–80, 141–142, 159–162, 203
  and scientific revolution, 130–136, 206–207
  See also criteria, aesthetic; induction, aesthetic
Cartesian natural philosophy, 56–58

Index

Index

Newton's theory of gravitation:
aesthetic appreciation for, 19–22, 111,
121
metaphysical allegiance of, 57–58
simplicity of, 97, 108, 111, 116–117, 121

objectivism about value, 31–32
occult qualities, 56
Osborne, Harold, 18
Osiander, Andreas, 171

Pais, Abraham, 186, 188
Palter, Robert, 167
Panofsky, Erwin, 180
particle physics, 51
Paxton, Joseph, 145–148
Penrose, Roger, 42, 84, 90
perception, aesthetic, 61–64
Perret, Auguste, 155–157
Pevsner, Nikolaus, 159
phenomena, natural:
aesthetic appreciation of, 19–21, 23
properties of, 41, 98–100, 106
unification of, 109–111
phlogiston theory, 49
physics. *See individual branches and theories*
physiology, 47–48
Planck, Max, 42, 116, 189, 196–197
Platonism, 30, 37, 115, 205
Poincaré, Henri, 22
Poisson, Siméon-Denis, 53
Polanyi, Michael, 43
Popper, Karl R., 71–72
positivism, 45, 50, 132, 200
logical, 13–16
precepts, methodological, 70–75
Pritchard, Thomas F., 142
Proclus, 169
projectivism about value, 31–32
properties, aesthetic, of scientific theories:
defined, 32–35
and phenomena, 98–100
recognition of, 35–38
surveyed, 39–60
transposition of, 28–29
See also abstractness; metaphysical allegiance; models, invocation of; simplicity; symmetry; visualization
Ptolemaic theory, 20, 165–176
Pythagoreanism, 171, 174, 205

quantum theory:
development of, 188–201
quantum electrodynamics, 49, 94–95
wave–particle dualism, 42–43, 193

Quine, Willard V. O., 131. *See also* Duhem-
Quine thesis

Ransome, Ernest L., 154
rationalist image of science, 7–12, 24,
202–207
realism, scientific, 10, 117
Reichenbach, Hans, 112
Reinhold, Erasmus, 172–173
relativity theory:
aesthetic appreciation for, 15–17, 93–94,
121, 187–188
artistic responses to, 21
development of, 183–188
simplicity of, 116–117, 121
Rennie, John, 143
representationalism in art theory, 68
representations of scientific theories, 24–
29, 84–85
Rescher, Nicholas, 73
revolution, scientific:
and aesthetic commitments, 128–133,
203–204, 206–207
and artistic revolution, 132–133
characteristics of, 125–126
Keplerian theory as, 180–181
Kuhn on, 8, 127–130, 134–139, 206
and political revolution, 139–140
quantum theory as, 194–195, 199–201
rationality of, 205–207
Rheticus, Georg Joachim, 173
Rickman, Thomas, 145
Rohrlich, Fritz, 17
Rosen, Joe, 65
Rosenfeld, Léon, 200
Ruskin, John, 144
Russell, Bertrand, 84–85
Rutherford, Ernest, 14, 45, 189

Schädlich, Christian, 151
Schrödinger, Erwin, 69, 190–193
Schwinger, Julian, 94
Sciama, Denis, 69
Scott, George Gilbert, 145
Semper, Gottfried, 144–145, 153
Shaftesbury, Lord, 62–63
Simonton, Dean K., 14
simplicity:
in biological science, 59–60, 117–118
of Copernican theory, 116, 167–168
of Newton's theory of gravitation, 97,
108, 111, 116–117, 121
in physical science, 59–60
in theory evaluation, 105–124

Index